Leonard Mascall

A Book of the Art and Manner, How to Plant and Grave all Sorts of Trees

Leonard Mascall

A Book of the Art and Manner, How to Plant and Grave all Sorts of Trees

ISBN/EAN: 9783744680912

Printed in Europe, USA, Canada, Australia, Japan

Cover: Foto ©berggeist007 / pixelio.de

More available books at **www.hansebooks.com**

"This famous book, with the exception of the Addition, is a translation of David Brossard's L'Art et manière de semir et faire pépinières."

Quaritch's Catalogue, no. 340, Dec. 1915, p. 223.

¶A BOOKE

of the Arte and maner how to Plant
and Graffe all sortes of Trees, how to set Stones
and sowe Peping, to make wilde Trees to
Graffe on, as also remedies and Medicines. With diuers o-
ther newe practises, by one of the Abbey of Sainct Vincent
in Fraunce, practised with his owne handes: deuided into
seuen Chapters, as hereafter more plainly shal appere
with an addition in the ende of this booke, of
certaine Dutche practises, set forthe
and Englished, by Leonard
Mascall.

In laudem incisionis distichon,
Hesperidum Campi quicquid Romanaque tellus,
Fructificat nobis, incisione datur.

¶Imprinted at London, for Jhon Wight.
Anno M.D.LXXXII.

The wight that willyng is to knowe,
The waie to Graffe and Plant:
Maie here finde plentie of that skill,
That erst hath been but scant.

To Plant or Graffe in other tymes,
As well as in the Spryng:
I teache by good experience,
To doe an easie thyng.

The pleasure of this thyng is greate,
The profite is not small,
To suche men as will practise it,
In thynges meere naturall.

The poore man maie with pleasure finde,
Some thyng to helpe his neede:
So maie the riche man reape some fruite,
Where carst he had but weede.

The noble man that needeth naught,
Maie thereby haue at will:
Suche pleasaunt fruite to serue his vse,
And giue eache man his fill.

The common weale can not but winne,
Where eache man doeth entende:
By skill to make the good fruites mo,
And ill fruites to amende.

Weigh well my wordes, and thou shalt finde,
All true that I doe tell:
Myne Aucthour doeth not write by gesse,
Practise made hym excell.

If thou wilt practise as he did,
Thou maiest finde out muche more:
He hath not founde out all the truthe,
That Nature hath in store.

Farewell.

¶ To the right Honourable, and my
very good lorde, Sir Jhon Paulet knight,
Lorde S. Jhon: Leonard Mascell wi-
sheth prosperous health, with con-
tinuall encrease of honor.

RIGHT *honorable, emõg
all Sciences that maie be lightly
obtained, & emong many goodlie exercises for menne, there is
none (emong the reste) more
mete & requisite, or that more
doeth refreshe the vitall spirites
of men, nor more engender ad-
miration in the effectes of Nature, or that is cause of greater recreation to the wearie and traueiled spirit of man, or
more profitable for mannes life, then is the skill of Planting
and Graffyng, the whiche not onely wee maie see with our
eyes, but also feele with our handes in the secrete woorkes of
Nature : yea, nothyng more discouereth vnto vs the greate
and incomprehensible woorke of GOD, that of one little Pepin seede, Nut, or small plant, maie come the self same herbe
or tree, & to bryng forthe the infinite of thesame fruite, whiche
also doeth shine and shewe forthe it self vnto vs, especially in
the Spryng tyme, by their diuersitie of shootes, blossoms, and
buddes, in diuers kindes of Nature, by the goodnesse and
mightie power of the greate Lorde and Creatour towardes
his people, in suche thynges as commeth forthe of the naturall yearth, to nourishe, to substaine, and maintaine our li-*

A .ii.   ues.

nes . *What greater pleasure can there bee, then to smell the sweete odour of Herbes, Trees, and Fruites, and to beholde the goodlie colour of the same , whiche in certaine tymes of the yere commeth forthe of the Wombe of their mother and nourse , and so to vnderstande the secrete operation of the same. And to bee short, of this labour (in our liues) wee doe take part thereof with great gaines and renenues that come thereby , whereas through idlenesse there commeth none: therefore to augment the same , it shall bee good to appeale and mitigate all fonde delightes, and vaine pleasures , with suche like vanities , and cleane put out and abolishe the delightes of all vices. Wherefore the Poet saieth: Let vs praise the true labouryng hower of the true labourer. Therevpon many greate Lordes and noble personages , haue lefte their Theaters, pleasaunt stages, goodly pastymes : forsaking and despisyng their pleasures , not muche regardyng riche Diademes , and costly parfumes , but haue giuen themselues to Plantyng and Graffyng, and suche like. In suche sorte , that if wee should diligently searche , and recite all the discourse of auncient Histories, as of late daies wee should finde, that the moste noble personages through their vertue , hath shewed many goodlie examples , as in one Theatre a supreme degree Honorable: nor haue had nothyng more deare, more requisite, nor more greatly in commendation, then Plantyng and Graffyng of fruite. Cyrus a greate Kyng of the Persians (as witnesse Xenephon,) did so muche delite in the Art of Plantyng and Graffyng, (whiche did shewe a great praise and glorie vnto his personage) that he had no greater desire or pleasure , then when he might. occupie hymself in Plantyng and Graffyng, to garnishe the yearth, to place and or=*

der

der, thereon certaine number of Trees. The Emperour Dio-
clesian, (as doeth recite, Sextus Aurelius Victor,) of his
owne good will without any constraint, did leaue the Scep-
ter of his Empire, for to remaine continually in the fieldes.
So muche pleasure did he take in Plantyng of fruite, in ma-
king of Orchardes and Gardens, whiche he did make, garnishe, and finishe with his owne handes.

  The Senatours, Dictatours, and Consuls of the Romaines, emong all other thynges haue commended Plantyng
and Graffyng, to bee one of the moste flourishyng labours in
this worlde for the Common wealthe, the whiche was cele-
brated and counted a greate vertue: yea, thei did so muche
esteeme it, that thei did hang Tables thereof in diuers places, neuer thinking the tyme more aptly spent, then in Plantyng and Graffyng, nothyng more contentyng themselues,
nothyng more delighted in any other affaires for the Common wealthe, then in Settyng, Sowyng, or Plantyng on the
yearth. How muche wee maie praise of late daies, and commende our Trauailers from other Countries, it is easie to
bee perceiued and knowne, but of Lordes, Gentlemen, and
Merchauntes, whiche haue had (as it doeth appere) a greate
regard in these latter daies, how thei might followe the exam-
ple of others: Whereby it hath replenished this our Realme
with diuers straunge Plantes, Herbes, and Trees, very good
and necessarie for the Common wealthe, not heretofore commonly knowne. And beholde, aboue all labours (for the common wealthe) wee ought to giue a sure and certaine iudge-
ment, that Plantyng and Graffyng. is more highly to bee
commended and praised, then many other worthie and noble thynges in this worlde: For this Arte hath not onely

The Epiſtle.

from tyme to tyme, been put in vſe and practiſe of labour,
through Kynges and Princes: But alſo it hath been put in
writyng of many greate and worthie perſonages, in diuerſe
kinde of languages, as in Greeke by Philometor, Hieron,
Acheleus, Orpheus, Muſceus, Homer, Hoſiode, Conſtan-
tine, Cæſar: And in Latine, by Verron, Caton, Columella,
Paladius, Virgill, Amilius Macer: and in the Portingall
tongue by Kyng Attalus and Mago, (the whiche reciteth
the Hiſtories) that after their death, the bookes of Planting and Graſſyng were brought to Rome, ſone after the de-
ſtruction of Carthage. Likewiſe how many ſince haue written onely of zeale, and loue for their Countrie and Common wealthe, of the fruitfull Arte of Plantyng and Graf-
ſyng: yea of late daies how many worthie men by their lear-
uyng haue written likewiſe thereof, ſhould ſeeme that it
hath come from their auncestours, as the greateſt honour,
through the noble inuention of the ſame. Likewiſe I dare
boldlie affirme, not onely the learned haue written, but
alſo haue been practiſers and inuenters of the ſame, (as wit-
neſſeth diuers Hiſtories) in diuers and many ſecrete thynges, wherein I dare boldlie ſaie, thei haue not learnedly,
nor ſo exactly written, but they haue more exactly taught
and left in writyng thynges ſo certaine, that their ſucceſſors
maie eaſily marke, obſerue, and keepe the ſame: For euery
one hath written, accordyng to the nature of his Countrie.
The Greekes for Greeke, the Barbarians for Barbarie, the
Italians for Italie, the Frenche menne for Fraunce, &c.
Whiche writyng without the order and practiſe, doeth very
ſmall profite for this our Realme of Englande, the whiche I
can blame nothyng more, then the negligence of our Nation,
whiche

The Epistle.

whiche hath had small care here tofore in Plantyng and
Graffyng, in some places of this Realme (as I haue knowne)
where as good and well disposed haue graffed , the euill and
malicious persone hath soone after destroied them againe:
but if wee would endeuour our selues therevnto (as other
Countries doe,) wee might florishe, and haue many a strange
kinde of fruite (whiche now wee haue oftentymes the want
thereof) that might greatly pleasure and serue many waies,
both the for riche & poore, as well as in Grece, Barbarie, I-
talie, or Fraunce, if our nation were giuen so well that waie,
as thei are.

   Right honourable , for so muche as I haue been long, in
declaryng of our predecessours, I will now leaue, (troublyng
your honour any further) and reste from the other thynges
accomplishyng my desire, besechyng your Lordshippe to take
this my simple trauaile in good part, requesting no other recompence for my paine herein , but wishyng it might bee put
in a generall practise through this Realme, whereby in smal
tyme would growe vnto a greate profite and commoditie,
bothe to the riche and poore, wherin I should thinke my self
not onely happie , but also to haue a iuste tribute for my de-
sertes , and also this Realme might thereby receiue no small
benefite, with praise of other Countries, in followyng our predecessours in this Arte of Plantyng and Graffyng : Some
places of this Realme are greatly commended and praised
emong others: as Kent for the cheefest, whiche vertue (notwithstandyng) can not bee cleane put out or forgotten, specially , when suche as your honour shall seeme to fauour the
same , and also to see the forward doynges thereof, in suche
groundes and Lordships as ye doe possesse, the whiche at this
                                                              tyme

The Epiſtle.

tyme hath onely moued me to attempte vnto your honour,
this my ſimple trauaile, whiche is not the onely duetie that
I owe vnto your honour, but as a due deſart, not thinking my
ſelf halfe able to recompence your vertuous liberalitie, nor
ſhewyng otherwaies how to recompence theſame, hath bold-
ned me at this tyme, to commende this my ſimple and rude
woorke vnto your Lordſhip, not accordyng vnto your eſtate
and honour, the whiche had been more meete and requiſite,
to haue had the finiſhyng of ſome better woorke. Therefore
beſechyng your honour, to weye and accept this myne intent
and good will herein, whiche thyng ſo doyng, I ſhall thinke
my ſelf not onely happie, but it ſhall encourage me the
more hereaſter, to take in hande the like or better woorke. Thus I beſeche the almigh-
tie GOD and Creatour, to en-
creaſe your proſperous honour, with long life
in health.

Entle Reader thou shalt vn=
derstande, J haue taken out of
diuers Authours this simple
worke,into our English tōgue,
praiyng thee for to accept it in
good parte: in so dooyng thou
shalt bolden me to trauaile further therin : and
thus shewyng my good will in declaring of di=
uers waies of Plantyng and Graffyng, and
how in the meetest tymes of the yere, with she=
wing of diuers cōmodities and secrets herein:
How to set or plāt with the roote,and without
the roote: how to sowe or sett Pepins or Cur=
nels,w̄ the ordering thereof. Also how to clense
your Graffes and Cions, how to helpe barren
and sicke trees : how to kill wormes & vermin,
and to preserue and keepe fruite : how to plant
and proyne your Uines , & to gather and presse
your Grape : how to cleanse and Mosse your
Trees: how to make your Cyder and Perrie:
how to set,choose,order and keepe Hops, with
many other secrete practises, whiche shall ap=
peare in the Table followyng, that euery per=
son may easily perceiue in these our daies more
largely of the Arte of Planting and Graffing,
thē heretofore hath been shewed. Which thyng
is not an exercise onely to the mynde , but like=
wise a greate profite many waies , with main=
tenance of health vnto the bodie . Therefore
spare not the bodie to shewe so great goodnesse

there=

therevnto, and alſo to the Common wealthe.
In theſe Daies(among the reſt)ye maie ſee ma=
ny, whiche be of the baſe and abiect ſorte of the
Common wealth,as thoſe which will not ſtick
to ſaie: fie on thee ſlaue. What thyng is now
counted moꝛe filthic in theſe Daies (amõg faire
perſonages)then labouring of the earth,which
we muſt all liue by. Well,theſe bee daintie per=
ſons:yet therevnto, what thyng is moꝛe beau=
tifull to the eye,moꝛe pꝛofitable to the purſſe, oꝛ
moꝛe healthfull vnto the bodie ? And herein to
put awaie all nouriſhyng of vice and idleneſſe,
it is caſie to recite infinite and many wooꝛthie
Loꝛdes and Gentlemen, whiche haue had a
greate care to followe the exãple of others.
Wherefoꝛe,Gentle Reader,let vs now leaue of
from all wanton games and idle paſtimes,and
bee no moꝛe as childꝛen whiche ſeeke but their
owne gaine and pleaſure,let vs therefoꝛe ſeeke
one of vs foꝛ an other in all good wooꝛkes foꝛ
the Common wealth,whereby thoſe that dooe
come after vs,maie ſo enioye our wooꝛkes and
trauaile herein, as we haue doen of our pꝛode=
ceſſoꝛs,that therein God maie be gloꝛified,
pꝛaiſed and honoꝛed in all our woꝛkes
of Plantyng and Graffyng: and
we therefoꝛe maie be thank=
full,from age to age,du=
ryng this moꝛtall.
life.Amen.

¶ The Table of all the principall thynges
contained in this booke, whiche ye shall
hereafter finde by nomber and leaf.

Of the seuen Chapters follcwyng.

¶ The first Chapter treateth of the settyng of Curnelles, of
Apple trees, Plum trree, Peare trees, and Seruice trees.

¶ The seconde Chapter treateth, how to set your wilde trees
come of Pepines, when thei be first plucked vp.

¶ The third Chapter, is of the settyng of Trees,
whiche come of Nuttes.

B.ij.　Of

The Table.

The

The Table.

The Table.

The Table.

¶ The seuenth Chapter is of medicinyng and keepyng
the Trees, when thei are planted and set.

To.

The Table.

To

The Table.

                          C.j.   To

The Table.

Praises be to God on hye,
In all our worldiy plantyng:
And let vs thanke the Romaines also,
For the Arte of Graffyng.

FINIS.

Lwaies before ye doe intende
to Plant or Graffe, it shall bee
meete to haue good experience
in thinges meete for this Arte,
as in knowyng the Natures
of all Trees and fruites, and
the differences of Climates, whiche be contrarie in euery lande: also to vnderstande the East
and West windes, with aspects and Starres,
to the ende ye maie beginne nothyng that the
winde or raine maie oppresse, that your labour
bee not lost, and to marke also and consider the
disposition of the Elementes that present yere,
for all yeres be not of like operation, nor yet after one sort: the Sommer and Winter doe not
beare one face on the yearth, nor the Spryng
tyme alwaies rainie ; or Autumne alwaies
moyst: of this none haue vnderstandyng, without a good and liuely markyng spirite, fewe or
none (without learnyng) maie discerne of the
varieties and qualities of the earth, and what
he doeth aske or refuse . Therefore it shall bee
good to haue vnderstandyng of the grounde
where ye doe Plant, either Orcharde or Garden with fruite: first it behoueth to make a sure
defence, to the end, that not onely rude persons
and children maie bee kept out, but all kinde of
hurtfull Cattell indomaging your Plantes or
Trees, as Oxen, kine, Calues, Horse, Hogges
and Sheepe, as the rubbyng of Sheepe doeth

greatly burne the Sappe, and often doeth kill
yong Trees and Plantes, and where thei are
broken, or brused with Cattell, it is doubtfull
to growe after. It shall bee good also, to sette,
Plant, or Graffe Trees all of like nature, and
strength together, that the greate and high
Trees, maie not ouercome the low and weake
for when thei bee not like of heighth, thei grow
nor ripe not your fruite so well at one tyme, but
the one before the other: That yearth whiche
is good for Uines, is good also for other fruite.

Ye muste digge your holes a yere before ye
Plant, that the yearth maie bee the better sea=
soned, mortified, and waxe tender, bothe by
Raine in Winter, and Heate in Sommer, that
thereby your Plantes maie take roote the soo=
ner, if ye will make your holes, and plant bothe
in a yere, at the least, ye ought to make your ho=
les twoo Monethes before ye Plante, and as
soone as thei bee made, then it shall bee good to
burne of Strawe, or suche like therin, to make
your grounde warme: the furthe ye make them
a sonder, the better your Trees shall beare:
make your holes like vnto a Fornaice, that is,
more straight in the mouthe then beneathe,
whereby the rootes maie haue the more rome,
and by straightnesse of the mouthe, the lesse
Raine or colde shall enter by in Winter, and al=
so lesse Heate to the roote in Sommer. Looke
also that the yearth ye putte to the rootes, bee
neither wette, nor laied in water: Thei dooe
commonly

commonly leaue a good space betwixte euery
Tree,for the hangyng bowes , for beyng nigh
together, ye can not sette rootes, nor sowe no=
thyng so well vnder your Trees, nor thei will
not beare fruite so well : Some soweth fourtie
foote,some thirtie betweene euery Tree: Pour
Plantes ought to bee greater then the handell
of a Shouell,and the lesser the better: See thei
bee straight, without knottes, or knobbes, ha=
uyng a long straight graine or barke , whiche
shall the sooner bee apte to take Graffes , and
when ye sette braunches or boughes of olde
Trees,choose the yongest and straitest branch
thereof, and those Trees whiche haue borne
yerely good fruite before , take of those whiche
bee on the Sunnie side, sooner then those that
growe in the couerte or shadowe, and when ye
take vp or alter your Plantes , ye shall note,to
what windes your Plant is subiecte,and so let
theim bee sette againe, but those whiche haue
growne in drie groundes, let theim bee sette in
moiste grounde : Pour Plantes ought to bee
cutte of three foote long . If ye will sette twoo
or three Plantes together in a hole, ye muste
take heede the roote of one touche not one an
other,for then the one will perishe and rotte the
other,or die by Wormes or other Vermen,and
when you haue placed your Plantes in the
pearth , it shall bee good to strike doune to the
bottome of euery hole , twoo shorte stakes as
greate as your arme, on either side your hole

C,iii.   one,

one, and let them appeare but a little aboue the pearth, that ye maie (thereby in Sommer) giue water vnto the rootes if neede be. Your young Plantes, and rooted Trees are commonly set in Autumne, from the first vnto the fiftcene of October, yet some oppinion is, better after Alhallowtide vnto Christmas, then in the spryng, because the pearth will drie to soone after, and also to set Plants without roote after Michelmasse, that thei maie the better mollifie and gather roote against the Spryng, wherof ye shall finde hereafter more at large. Thus muche haue I thought meete to declare vnto the Planters and Graffers, whereby thei maie the better auoide the occasion and daungers of Plantyng and Graffing, whiche maie come often tymes through ignoraunce.

1575.

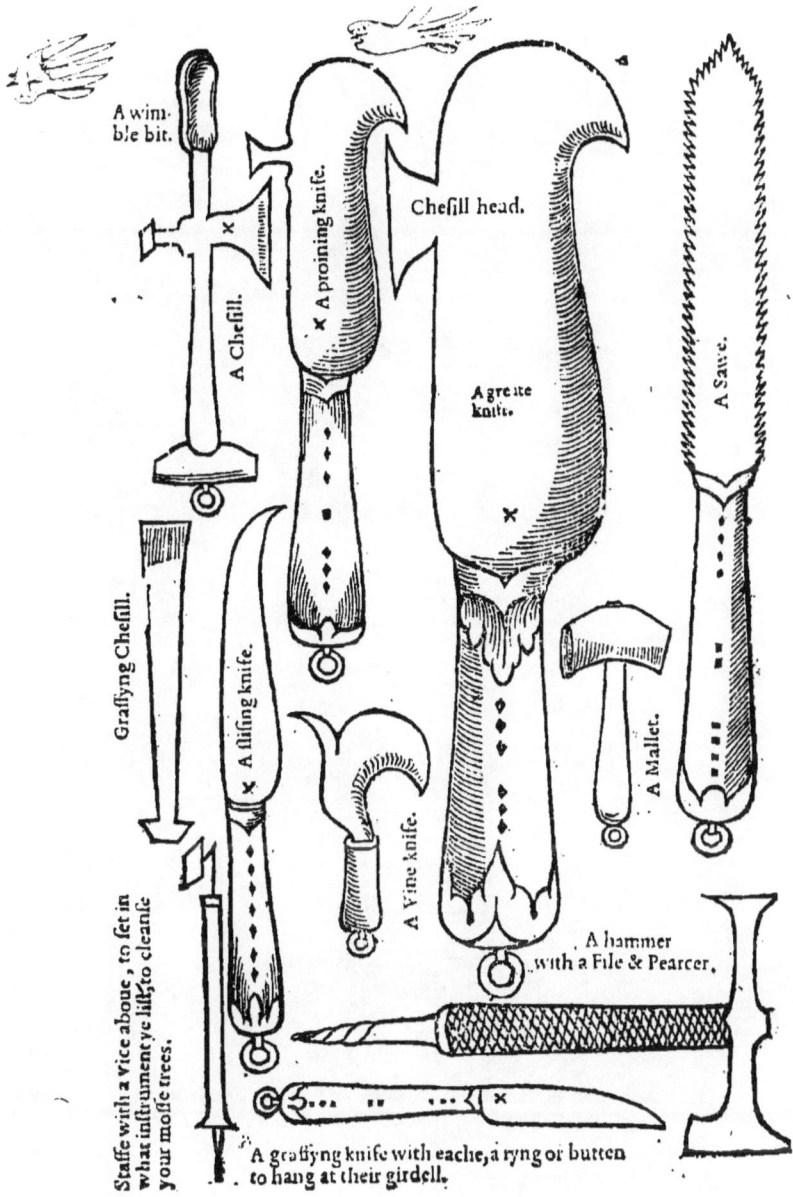

A wim-
ble bit.

A proining knife.

Cheſill head.

A Cheſill.

A greate
knife.

A Sawe.

Graffyng Cheſill.

A ſliſing knife.

A Vine knife.

A Mallet.

Staffe with a vice aboue, to ſet in
what inſtrument ye liſt, to cleanſe
your moſſe trees.

A hammer
with a File & Pearcer.

A graffyng knife with eache, a ryng or butten
to hang at their girdell.

¶ The Arte of Plantyng

and Graffyng:

The first Chapter.

¶ This Chapter treateth of the settyng of
Curnels, young Plum trees and Pears trees,
of Damsons and Seruice trees.

OR to make young Trees of the Pepins, of
Apples, Peares, Plummes and Seruice. Firſt
ye muſt prepare and make a great bed or quarter well repleniſhed, blende or mixt with good
fatt pearth, and placed well in the Sunne, and
to bee well laboured and digged a good tyme before you doe
occupie it: and if ye can by any meanes, let it bee digged very
deepe the Winter before, in blindyng or mixing it well together with good fatt pearth, or els to bee mixed almoſt the half
with good dung : and ſo let it rette and ripe together with the
pearth. And ſee alwaies that plott be cleane vnto the preſſyng
of Spoer, that no wilde Cions or Plants doe ſpring or grow
thercon. Then in the Moneth of September, December, or
there aboutes, take of the Pepins, or Pomes of the ſayd fruite
at the firſt preſſyng out of your licour, before the Curnels be
marred or bruiſed: then take out of them, and rubbe a fewe at
once in a cloth, and drie them betwixt your handes, and take
ſo many thereof as you ſhall thinke good: then make your bed
ſquare, faire and plaine, and ſowe your ſeedes thereon, then
take and couer them with a Rake lightly, or with pearth, not
puttyng to muche vpon them. This doen, deuide your beddes
into quadrantes or ſquares of foure foote brode or thereabout,
that when ye liſt ye may cleanſe them from the one ſide to the
other, without treadyng thereon . Then ſhall ye couer your
Seedes or Pepins with fine pearth, ſo ſiftyng all ouer them,
that then thei maie take the deeper and ſurer roote, and will
keepe them the better in Winter followyng , and if ye liſt ye
maie rake them a little all ouer : ſo that ye raiſe not your Pepins aboue the pearth.

D.   An

¶ An other waie how one maie take the Pepins at the
first commyng of the licour or pressyng.

Whiche is: ye shall choose the greatest & fairest Curnels or Pepins, and take them forth at the first brusing of your fruite, then drie them with a cloth, and
keepe them all the Winter vntill S. Androwes tide: then a
little after sowe them in good yearth, as thinne as ye doe sow
Peason, and then rake them ouer as the other.

¶ How one ought to vse his yearth to sowe
Pepins without dunging.

But in this maner of doyng (in the Spring) it is not so
greate neede for to raise or digge the yearth so deepe as
that whiche is dunged in Winter: but to deuide your
quarters, in coueryng your Pepins not so muche with earth
as those whiche be sowne with good dung, but when ye haue
sowne them, a little rake them all ouer.

¶ How ye ought to take heede of Poultrie for
scrapyng of your beddes or quarters.

As sone after as your Pepins bee sowne vppon your
beddes or quarters, let this be doen, one waie or other,
that is, take good heede that your Hennes doe not
scrape your beddes or quarters: therefore sticke them all ouer
light and thinne with bowes, or thornes, and take good heede
also to Swine, and other Cattell.

¶ How to weede or cleanse your beddes
and quarters.

And when the Winter is past and gone, and that ye see
your Pepins rise and growe: so lett them increase the
space of one yere, but see to cleanse weedes, or other
thynges whiche maie hurte them, as ye shall see cause. And in
the Sommer when it shall waxe drie, water them hardly in
the Euenynges.

¶ How one ought to plucke vp the wilde Cions.

And when these wilde Cions shall be great, as of the
growth of one yere, ye must then plucke them vp all in
Winter folowyng, before thei doe beginne to spryng.
againe.

againe. Then shall ye set them and make of them a wild Oꝛ
charde as followeth.

*¶ The seconde Chapter treateth how one
shall set again the small wilde trees, which come
of Pepins, when thei be first pluckt vp.*

OR the Bastarde oꝛ little wilde Trees inconti
nent assone as thei be pluckt vp, ye must haue of
other good yearth well trimmed and dunged and
to bee well in the Sunne, and well pꝛepared and
dꝛest, as it is saied in the other parte before of the Pepins.

*¶ How to dung your Bastarde or wilde yong
trees whiche come of Pepins.*

ABout Aduent before Chꝛistmas, ye must digge and
dung well the place where as ye will set them, & make
your square of yearth euen and plaine, so large as ye
shall thinke good, then set your wilde trees so farre one from
an other as ye thinke meete to bee grafr, so that thei maie bee
set in euen rankes and in good oꝛder, that when neede shall re
quire, ye maie remoue oꝛ renue any of thē oꝛ any part therof.

*¶ How ye ought in replantyng or settyng, to cut of in
the middest the principall greate rootes.*

IN what parte soeuer ye doe set any Trees, ye must cut of
the greate maister roote, within a foote of the stocke, and
all other bigge rootes, so that ye leaue a foote long thereof,
and so let them be sett, and make your rankes crossewise one
from an other halfe a foote, oꝛ there aboutes, and ye must also
see that there be of good dung moꝛe deepe and lower then ye
doe set your Trees, to comfoꝛte the saied rootes withall.

¶ How you ought to set your Trees in rancke.

YE shall set your small yong Trees in rankes, halfe a
large foote one from an other: and let them be couered
as ye doe set thē, with good fat earth al ouer the rootes.

¶ How to make the space from one rancke to an other.

YE shall leaue betweene your rankes, from one rancke
to an other, one foote, oꝛ there aboutes, so that ye maie

D.ij. passe

passe betweene euery rancke for to clense them if neede require, and also for to graffe any parte or parcell thereof when
tyme shall be meete. But ye must note, in makyng thus your
ranckes, ye shall make as many allies as ranckes. And if ye
thinke it not good to make so many allies, then deuide those
into quarters of fiue foote broade, or thereaboutes, and make
and set foure ranckes (in eche quarter of the same) one foote
from an other, as ye vse to set greate Cabbage. And assone after as ye haue set them in ranckes and in good order as is a-
foresaied, then shall ye cut of all the Sets euen by the ground.
But in thus doyng, see that ye doe not plucke vp or loose the
pearth which is about them: or if ye will, ye maie cut them before ye doe set them in ranckes. If ye doe so, see that ye set
them in suche good order, and euen with the earth, as is afore
sayd. And it shall suffice also to make your ranckes as ye shall
see cause. And looke that ye furnishe the pearth all ouer with
good dung, without minglyng of it in the pearth, nor yet to
couer the saied Plantes withall, but strowed betwixt: and ye
must also looke wel to the cleansing of weedes, grasse, or other
such thinges which wilbe a hurt to the growth of the Plantes.

¶ How to water Plantes when thei waxe drie.

IT shall bee good to water them when the tyme is drie: in
the first yere. Then when thei haue put forth of new Cions,
leaue no more growyng but that Cion whiche is the principall and fairest, vppon euery stocke one: all the other cut of
harde by the stock: and euer as there doe grow small twigges
about the stocke, ye shall (in the Moneth of Marche and A-
prill) cut them all of harde by the stocke. And if ye then sticke
by euery Plant a pretie wande, and so binde them with Willowe barke, Brier, or Osiers, it shall profite them muche in
their growth. Then after fiue or sixe yeres growth, when thei
bee so bigge as your fingar, or there aboutes, ye maie then remoue any of them whereas ye will haue them growe and remaine.

¶ How one ought to remoue Trees,
and to plant them againe.

The

The maner how ye ought to remoue Trees, is shewed in the sixt Chapter following: then about twoo oʒ thʒe yeres after their remouyng, ye shall Graffe them, foʒ then thei will bee the better rooted. As foʒ the others whiche ye leaue still in ranckes, ye maie alsо graffe them where as thei stande, as ye shall see cause good. When ye haue plucked vp the fairest to Plant in other places (as is afoʒesaied) also the maner how to Graffe them, is shewed in the sixt Chapter followyng. But after thei shall bee so grafte, in what place so euer it bee, ye shall not remoue oʒ set them in other places a‐ gaine, vntill the Graffes bee well closed vpon the hedde of the wilde stocke.

¶ When the best tyme is to replant, or remoue,

Hen the hedde of the stocke shall bee all ouer closed about the graffes, then ye maie when ye will, tranſ‐ plant and remoue them (at a due tyme) where theï shall continue. Foʒ with often remouyng, ye shall doe them greate hurte in their rootes, and bee in daunger to make them die.

¶ Of negligence and forgetfulnesse.

IF peraduenture ye foʒget (thʒough negligence) and haue let small Cions two oʒ thʒee yeres grow about the rootes of your stockes vnplucked vp, then if ye haue so doen, ye maie well plucke them vp and set them in ranckes, as the o‐ ther of the Pepins. But ye must set the ranckes moʒe larger that thei maie bee remoued without hurtyng of eche others rootes: and cut of all the small twigges aboue as neede shall require, though thei bee set oʒ graffed. Oʒder them also in all thynges as those small Cions of a yeres growth.

¶ It is not so conuenient to Graffe the
Seruice Tree, as to set hym.

Hereas ye shall see yong Seruice Trees, it shall be moſt pʒofite in settyng them, foʒ if ye doe graffe them, I beleeue ye shall winne nothyng thereby. The best is onely to plucke vp the yong Bastard trees when thei are as greate as a good walkyng Staffe: then pʒoyne oʒ cut

cut of their braunches and carie them to set whereas thei may
bee no moze remooued: and thei shall profite moze in setting
then graffing.

¶ Some Trees without Graffing bryng forth good
fruite, and some other beyng graffed bee
better to make Syder of.

IT is here to be marked, that though the Pepins be sowne
of the Pomes of Peares and good Apples:yet ye shal finde
that some of them doe loue the Tree whereof thei came:
and those bee right, whiche haue also a smooth barke, and as
faire as those whiche bee graffed: the which if ye plant oz set
them thus growing from the maister roote without graffing,
thei shal bzing as good fruite, euen like vnto the Pepin wher-
ofhe first came . But there bee other newe sortes commonly
good to eate , whiche be as good to make Syder of, as those
whiche shall be graffed foz that purpose.

¶ When you list to augment and multiplie your Trees.

AFter this sorte ye maie multiplie them , beyng of diuers sortes and diuersities, as of Peares oz Apples,oz
suche like . Notwithstandyng , when soeuer you shall
finde a good Tree thus come ofthe Pepin,as is afozesaied, so
shall ye vse hym. But if ye will augment Trees of them selues, ye must take Graffes, and so graffe them.

¶ Of the maner and chaungyng of the
fruite of the Pepin Tree.

WHen soeuer ye doe replant oz chaunge your Pepin
Trees from place to place . in so remouyng often
the stocke,the fruite thereof shall also chaunge:but
fruite whiche doeth come of Graffyng , doeth alwaies keepe
the forme and nature ofthe Tree whereof he is taken: foz as
I haue saied,as often as the Pepin trees be remoued to a better grounde,the fruite thereof shall be so muche amended.

¶ How one ought to make good Syder.

HEre is to bee noted if ye will make good Syder of
what fruite soeuer it be, beyng Peares oz Apples,but
specially of good Apples , and wilde fruite , haue alwaies

waits a regarde vnto the ripyng thereof, so gathered dꝛie,
then put them in dꝛie places, on bourdes in heapes, couered
with dꝛie ſtrawe, and whenſoeuer ye will make Spꝺer there=
of, chooſe out all thoſe whiche are blacke bꝛuſes, and rotten
Apples, and thꝛowe them awaie, then take and vſe the reſt
foꝛ Spꝺer: But here to giue you vnderſtandyng, doe not as
thei doe in the Countrie of Mens, whiche doe put their fruite
gathered, into the middeſt of their Garden, in the raine and
miſtynges vpon the beare pearth, whiche will make them to
leeſe their foꝛce and vertue, and doeth make them alſo wiche=
red and tough, and lightly a man ſhall neuer make good Sp=
der that ſhall come to any purpoſe oꝛ good pꝛofite thereof.

¶ To make an Orcharde in fews yeres.

Some doe take yong ſtraight ſlippes, whiche doe growe
from the rootes, oꝛ of the ſides of the Apple trees, about
Michaelmaſſe, and doe ſo plant oꝛ ſet them (with Otes)
in good grounde, whereas thei ſhall not bee remoued, and ſo
graffe (beyng well rooted) thereon. Otherſome doe take and
ſet them in the Spꝛyng tyme (after Chꝛiſtmas) in likewiſe,
and doe graffe thereon when thei bee well rooted: and bothe
doe ſpꝛyng well. And this maner of waie is counted to haue
an Oꝛcharde the ſoneſt. But theſe Trees will not endure paſt
twentie oꝛ thirtie yeres.

¶ The third Chapter is of ſettyng Trees of Nuttes.

¶ How one ought to ſet Trees which come of Nuttes.

Oꝛ to ſet Trees whiche come of Nuttes: when
ye haue eaten the fruite, looke that ye keepe the
Stones and Curnels thereof, then let them be
dꝛied in the winde, without the vehemencie of
the Sunne, ſo reſerue them in a boꝛe, and vſe
them as befoꝛe.

¶ Of the tyme when ye ought to plant or ſet them.

YE ſhall plant oꝛ ſet them in the beginnyng of Winter, oꝛ
afoꝛe Michaelmaſſe, whereby thei maie the ſoner ſpꝛyng
out

out of the pearth. But this maner of setting is daungerous: for the Winter then comming in, and thei being young and tender in comming vp, the colde will kill them. Therefore it shall be best to staie and reserue them till after Winter. And then before pe doe sett them, ye shall soke or steepe them in Milke, or in Milke and Water, so long till thei doe stincke therein: then shall ye drie them and set them in good pearth in the chaunge or increase of the Moone, with the small ende vpwarde, fower fingers deepe, then put some sticke thereby to marke the place.

¶ For to set them in the Spryng tyme.

IF ye will plant or set your Nuttes in the Spryng tyme, where ye will haue them still to remaine and net to bee remoued, the best and most easie waie is, to set in euery suche place (as ye thinke good) thre or fower Nuttes nigh together, and when thei doe all spryng vp, leaue none standyng but the fairest.

¶ Of the dungyng and deepe digging thereof.

ALso whereas ye shall thinke good, ye maie plant or set all your Nuttes in one square or quarter togethers in good pearth, and dunged in suche place and tyme as thei vse to plant. But see that it be well dunged, and also digged good and deepe, and to be well medled with good dung throughout, then set your Nuttes thre fingers deepe in the pearth, and halfe a foote one from an other: ye shall water them often in the Sommer when there is drie weather, and see to weede them, and digge it as ye shall see neede.

¶ Of Nuttes and Stones like to the
Trees they came of.

IT is here to be noted, that certaine kinde of Nuttes and Curnels which doe loue the Trees, whereof the fruite is like vnto the Tree thei came of, when thei be planted in good grounde, and set well in the Sunne, which be: the Walnuts, Chestnuts, all kinde of Peaches, Figges, Almondes, and Abrycotes, all these doe loue the Trees thei came of.

¶ of

Of the Plantyng the saied Nuttes in good yearth, and in the Sunne.

ALL the saied Trees doe bzyng as good fruite of the saied Nuttes, if thei bee well Planted, and set in good pearth, and well in the Sunne, as the fruite and trees thei first came of.

Why fruite shall not haue so good sauour,

FOR if ye Plant good Nuttes, good Peaches, oz Fig-ges in a garden full of shadow, the which hath afoze lo-ued the Sunne, as the Uine doth, foz lack thereof, their fruite shall not haue so good sauour, although it bee all of one fruite: and likewise so it is with all other fruite and Trces, foz the goodnesse of the pearth, and the faire Sunne, doeth pze-serue them muche.

For to set the Pine tree.

FOR to sett the Pine tree, ye must sett oz Plant them of Nuttes, in Marche, oz about the shoote of the sappe, not lightly after, ye must also set them where thei maie not bee remoued after, in holes well digged, and well dunged, not to be transplanted oz remoued againe, foz very hardly thei will shoote fozth Cions, beyng remoued, specially if ye hurte the maister roote thereof.

For to set Cherrie trees.

FOR to set sowze Cherries whiche doe grow common-ly in Gardens, ye shall vnderstande thei maie well grow of stones, but better it shall be to take of the small Cions whiche doe come from the great rootes: then plant them, and sooner shall thei growe then the stones, and those Cions must bee set when thei are small, young and tender: as of twoo, oz thzee yeares growth, foz when thei are greate, thei pzofite not so well: and when ye set them, ye must see to cut of all the bowes.

Trees of Bastarde and wilde Nuttes.

THere be other sortes of Nuttes, although thei bee well set in good ground, and also in the Sunne, yet will thei not bzyng halfe so good fruite as the other, noz com-

C.j.   monly

monly like vnto thofe Nuttes thei came of, but to bee a ba-
ftarde wilde fowꝛe fruite, which is the Filbert, fmall Nuttes,
of Plummes, of Cherries, and the great Abꝛꝑcots:therefoꝛe
if ye will haue them good fruite, ye muft fett them in maner
and foꝛme followyng.

How to fet Filberdes or Hafell trees,

FOR to fet Filbirdes oꝛ Hafels, and to haue them good,
take the fmall wandes that growe out from the roote of
the Filberde oꝛ Hafell tree (with ſhoꝛt hearie twigges)
and fet them, and thei ſhall bꝛyng as good fruite as the Tree
thei came of: it ſhall not bee needefull to pꝛoyne, oꝛ cut of the
bꝛaunches thereof when ye fet them, if thei be not greate: but
thofe that ye doe fet, let them bee but of twoo oꝛ thꝛee peares
growth, and if ye ſhall fee thofe Cions whiche ye haue planted, not to bee faire and good, oꝛ doe growe and pꝛofper not
wel, then (in the Spꝛyng tyme) cut them of hard by the roote,
that other fmall Cions maie growe thereof.

¶ To fet Damfons or Plum trees.

IN fettyng Damfons oꝛ Plum Trees, whiche fruite ye
would haue like to the trees thei came of: if the faied trees
bee not grafte befoꝛe, ye ſhall take onely the Cions that
growe from the roote (of the old ſtocke) whiche goweth with
fmall twigges, and plant oꝛ fet them: and their fruite ſhall bee
like vnto the Trees thei were taken of.

¶ To take Plum Graffes, and to Graffe them on other Plum Trees.

AND if your Plum trees bee graft alreadie, and haue
the like fruite that you defire, ye maie take your graffes thereof, and graffe them on your Plum trees, and
the fruite that ſhall come thereof, ſhall be as good as the fruit
of the Cion, whiche is taken from the roote, becaufe thei are
muche of like effecte.

¶ To fet all fortes of Cheries.

TO fet all foꝛtes of greate Cheries, and others: ye
muſte haue the graffes of the fame trees, and graffe
them on other Cherie trees, although thei be of a fower

wer fruite, and when thei are so graft, thei will be as good as
the fruite of the Tree whereof the graffe was taken : for the
stones are good, but to set to make wilde Cions, or Plantes
to graffe on.

¶ The maner how one maie order bothe
Plum trees, and Cherie trees.

FOr so muche as these are twoo kinde of trees, that
is, to vnderstande, the Cherie, and the Plum tree,
for when thei be so graft, their rootes be not so good,
nor so free as the braunches aboue, wherefore the Cions that
doe come from the rootes, shall not make so good and franke
trees of. It is therefore to be vnderstoode, how the maner and
sort is to make franke trees, that maie put forthe good Cions
in tyme to come, which is: when thei be greate and good, then
if ye will take those Cions, or yong spryngcs from the rootes
ye maie make good trees thereof, and then it shall not neede to
graffe them any more after: but to augment one by the other,
as ye doe the Cions from the roote of the Nutte, as is aforesaied, and ye shall doe as followeth.

¶ How to graffe Plum trees and Cherie trees.

YE maie well graffe Plum trees, and greate Cherie
trees, in suche good order as ye luste to haue them,
and as hereafter shall bee declared in the fifth Chapter following: for these would bee graffed while thei are yong
and small, and also grafte in the grounde, for thereby one maie
dresse and trim them the better, and put but one Graffe in eche
stocke of the same. Cleaue not the harte, but a little on the one
side, nor yet deepe, or long open.

How ye must proine or cut your Trees.

FOr when your Graffes be well taken on the stocke, and
that the Graffes doe put forthe faire and long, about one
yeres gouth, ye must proine, or cut the braunche of commonly in Winter, (when thei proine their Uines) a foote lower, to make them spread the better : then shall ye medole all
through with good fatte pearth, the whiche will drawe the
better to the place, whiche ye haue so proined or cut.

C.ij.   The

The conuenienst waie to clense and proine,
or dreſſe the rootes of trees.

AND for the better clensyng and proinyng trees beneath, is thus: ye ſhall take awaie all the weedes, and
graſſe about the rootes, then ſhall ye digge them ſo
rounde about, as ye would ſeme to plucke them. vp, and ſhall
make them halfe bare, then ſhall ye enlarge the yearth aboute
the rootes, and where as ye ſhall ſee them growe faire and
long, place or couche them in the ſaied hole and yearth again:
then ſhall ye put the cutte ende of the tree where he is graſte,
ſomewhat more lower then his rootes were, whereby his Cions ſo graft, ſhall ſprong ſo muche the better.

When the ſtocke is greater then the graſſes.

When as the Tree waxeth, and ſwelleth greater beneath the graffyng, then aboue: then ſhall ye cleaue
the rootes beneath, and wreathe them rounde, and ſo
couer them againe. But ſee ye breake no roote thereof, ſo will
he come to perfection. But moſte men doe vſe this waie: if the
ſtocke waxe greater then the graffes, thei doe ſlitte doune the
barke of the graffes aboue, in twoo or three partes, or as thei
ſhall ſee cauſe thereof: and ſo likewiſe, if the graffes waxe
greater aboue then the ſtocke, ye ſhall ſlitte doune the ſtocke
accordyngly, with the edge of a ſharpe knife. This maie well
be doen at any tyme in Marche, Aprill, and Maie, in the creaſe
of the Moone, and not lightly after.

The remedie when any bough or mem-
ber of a tree is broken.

IF ye ſhall chaunce to haue boughes, or members of trees
broken, the beſt remedie ſhall bee, to place thoſe bowes or
members right ſone again, (then ſhall ye comfort the rootes with good newe yearth) and bind faſt thoſe broken bowes
or members, bothe aboue and beneath, and ſo let them remain
vnto an other yere, till thei maie cloſe & put forth of new cions.

When a member, or bough is not bro-
ken, how to proine them.

Where

Where as ye ſhall ſee vnder oʒ aboue ſuperſluous bowes, ye maie cutte oʒ pʒoine of, (as ye ſhall ſee
cauſe:)all ſuche bowes harde by the Tree, at a due
tyme, in the Winter followyng. But leaue all the pʒincipall
bʒanches, and where as any are bʒoken, let them bee cutte of
beneath,oʒ els by the grounde,and caſt them awaie:thus muſt
ye doe yerely, oʒ as ye ſhall ſee cauſe, if ye will keepe your
Trees well and faire.

How one ought to enlarge the hole
about the Tree rootes.

IN pʒoinyng your Trees,ifthere be many rootes,
ye muſt enlarge them in the hole, and ſo to wʒeathe
them, as is afoʒeſaicd, and to vſe the without bʒeakyng,then couer them agains with good fat pearth
whiche ye ſhall mingle in the ſaied hole, and it ſhall bee beſt
to bee digged all ouer a little befoʒe,and ſee that no bʒanche oʒ
roote bee lefte vncouered,and when ye haue thus dʒeſſed your
Trees,if any roote ſhall put foʒthe,oʒ ſpʒing hereafter out of
the ſaied holes,in growyng,ye maie ſo pʒoine them as ye ſhal
ſee cauſe, in lettyng them ſo remaine twoo oʒ thʒee yeres after, vnto ſuche tyme as the ſaied Graffes bee ſpʒong vp, and
well bʒaunched.

How to ſet ſmall ſtaues by, to
ſtrengthen your Cions.

TO auoide daunger,ye ſhall ſet oʒ pʒicke ſmall ſtaues
aboute your Cions, foʒ feare of bʒeakyng, and then
after thʒee oʒ fower yeres, when thei bee well bʒaun-
ched:ye maie then ſet oʒ plant them in good pearth, (at the be-
gimyng of Winter) but ſee that ye cutte of all their ſmall
bʒaunches harde by the ſtocke,then ye maie plant them where
ye thinke good,ſo as thei maie remaine.

In taking vp Trees,note.

YE maie well leaue the maiſter roote in the hole (when
ye digge hym vp) if the remoued place bee good foʒ
hym, cutte of the maiſter rootes by the ſtubbe,but pare
not of all the ſmall rootes, and ſo plant hym, and he ſhall pʒo-

fite moꝛe thus, then others with all their maiſter rootes. Whē
as Trees be greate, thei muſt be diſbꝛaunched, oꝛ bowes cutt
of, befoꝛe thei be ſet againe, oꝛ els thei will hardly pꝛoſpere. Jf
the Trees bee greate, hauyng greate bꝛaunches oꝛ bowes,
when ye ſhall digge them vp, ye muſt diſbꝛaunche them afoꝛe
ye ſett them againe, foꝛ when Trees ſhall bee thus pꝛoined,
thei ſhall bꝛyng greate Cions from their rootes, whiche ſhall
bee franke and good to replant, oꝛ ſet in other places, and ſhall
haue alſo good bꝛaunches and rootes, ſo that after it ſhall not
neede to graffe them any moꝛe, but ſhall continue one after an
other to be free and good.

How to couche the rootes when thei are proined.

JN ſettyng your Trees againe, if ye will dꝛeſſe the rootes of ſuche as ye haue pꝛoined, oꝛ cutte of the bꝛaunches
befoꝛe, ye ſhall leaue all ſuche ſmall rootes whiche growe
on the great roote, and ye ſhall ſo place thoſe rootes in replantyng againe not deepe in the yearth, ſo that thei maie ſone
grow, and put foꝛthe Cions: whiche beyng well vſed, ye maie
haue fruite ſo good as the other afoꝛe mencioned, beyng of thꝛee
oꝛ fower yeres grouth, as afoꝛe is declared.

What Trees to proine.

THis waie of pꝛoinyng is moꝛe harder foꝛ the greate
Cherie (called Healmier) then foꝛ the Plum Tree.
Alſo it is verie requiſite and meete foꝛ thoſe Cions oꝛ
Trees, whiche be graft on the wilde ſowꝛe Cherie Tree, to
be pꝛoined alſo, foꝛ diuers and ſondꝛie cauſes.

Why the ſower Cherie dureth not ſo long, as the Healmier or greate Cherie.

THE wilde and ſowꝛe Cherie, of his owne nature will
not ſo long tyme endure, (as the greate Healme Cherie) neither can haue ſufficient ſappe to nouriſhe the
graffes, as the greate Healme Cherie is graft, therefoꝛe whē
ye haue pꝛoined the bꝛaunches beneath, and the rootes alſo, ſo
that ye leaue rootes ſufficient to nouriſhe the Tree, then ſett
hym. Jf ye cutte not of the vnder rootes, the Tree will pꝛofite

fite moze eafier, and alfo lighter to be knowne, when thei put
fozthe Cions, from the roote of the fame, the whiche ye maie
take hereafter.

To graffe one greate Cherie vpon an other.

YE muft haue refpect vnto the Healme Cherie, whiche
is graft on the wilde Goinire (whiche is an other kind
of greate Cherie) and whether you doe pzoine them
oz not, it is not materiall: foz thei dure a long tyme. But ye
muft fee to take awaie the Cions, that doe growe from the
roote of the wilde Goinire, oz wilde Plum Tree: becaufe thei
are of Nature wilde, and dooe dzawe the fappe from the faied
Tree.

Of deepe fettyng, or fhalowe.

TO fet your ftockes oz Trees fomewhat deeper on the
high groundes, then in the valleies, becaufe the Sun
(in Sommer) fhall not dzie the roote: and in the lowe
grounde moze fhallowe, becaufe the water in Winter fhall
not dzoune oz annoie the rootes. Some doe marke the ftocke
in takyng it vp, and to fet him again thefame waie, becaufe he
will not alter his nature: fo likewife the graffes in graffyng.

The fourth Chapter doeth fhewe how to
*fet other Trees which come of wilde Cions pricked
in the yearth without rootes: and alfo of
preynyng the meaner Cions.*

Trees taking roote pricks of braunches.

There bee certaine whiche take roote, beeyng
pzicked of bzaunches pzopned of other Trees,
whiche bee, the Mulberie, the Figge tree, the
Quince tree, the Seruice tree, the Pomegra-
nad tree, the Apple tree, the Damfon tree, and
diuers fortes of other Plum trees, as the Plum tree of Pa-
radife, &c.

How one ought to fet them,

foz

For to set these sorts of Trees, ye must cut of the Cions, twigges or boughes, betwixt Alhallowtide and Christmasse, not lightly after. Ye shall choose them whiche bée as great as a little Staffe or more, and looke whereas ye can finde them faire, smooth, and straight, and full of sappe withall, growyng of yong Trees, as of the age of three or fower yeres growth, or thereaboutes, and looke that ye take them so from the Tree with a brode Chisell, that ye breake not or loose any parte of the barke thereof, more then halfe a foote beneath, neither of one side or other: then proyne or cut of the braunches, and pricke them one foote deepe in the yearth, wel digged and ordered before.

How to binde them that be weake.

Those Plants whiche be slender, ye must proyne or cut of the braunches, then binde them to some stake or suche like to be set in good yearth, and well meddled with good dung, and also to be well and deepely digged, and to be set in a moyst place, or els to be well watred in Sommer.

How one ought to digge the yearth for to set them in.

And when that ye would set them in the pearth, ye must first prepare to digge it, and dung it well throughout a large foote deepe in the pearth. And when as ye will set them every one in his place made (before) with a crowe of Iron, and for to make them take roote the better, ye shall put with your Plantes, or watered Otes, or Barley; and so ye shall let them growe the space of three or fower yeares, or when thei shall bee well braunched, then ye maie remoue them, and if ye breake of the olde stubbie roote and set them lower, thei will last a long tyme the more. If some of those Plantes doe chaunce to put forthe the Cions from the roote, and bring so rooted, ye must plucke them vp though thei be tender, and set them in other places.

Of Cions without rootes.

If

ƒ that the faied Plants haue of Cions without roo=
tes, but whiche come from the Tree roote beneath,
then cut them not of till thei be of two or three yeres
grovvth, by that tyme, thei will gather of rootes to bee re=
planted in other places.

To plant the Figge tree.

HE faied Plantes taken of Figge trees graffed, be
the beſt: ye maie likewiſe take other ſortes of Figge
trees, and graffe one vpon the other, for like as vpon
the wilde Trees doe come the Pepins, euen ſo the Figge,
but not ſo ſoone to proſper and grovve.

How to ſet Quinces.

LJkewiſe the nature of Quinces is to ſpryng, if thei bee
pricked (as aforeſaied) in the yearth, but ſometymes I
haue graffed with great difficultie (faieth myne Authoꝛ)
vppon a white Thoꝛne, and it hath taken and boꝛne fruite to
looke on, faire, but in taſte moꝛe weaker then the other.

The waie to ſet Mulberies.

Here is alſo an other waie to ſett Mulberies as
folloveth, which is, if you doe cut in Winter certaine greate Mulberie bovves oꝛ ſtockes, aſunder
in the bodie (with a Savve) in troncheons a foote
long oꝛ moꝛe, then ye ſhall make a greate furrovve in good
yearth well and deepe, ſo that ye maie couer well againe your
troncheons, in ſettyng them an ende halfe a foote one from an
other, then couer them againe, that the yearth maie bee aboue
thoſe endes, three oꝛ fovver fingers high, ſo let them remaine,
and water them (in Sommer) if neede bee ſometymes, and
cleanſe them from all hurtfull weedes and rootes.

Note of the ſame.

Hat then within a ſpace of tyme after, the ſayd troncheons will put fooꝛth Cions, the whiche when thei
be ſomewhat ſprigged, hauyng two oꝛ three ſmall
twigges, then ye maie tranſplant oꝛ remoue them where ye
liſt: but leaue your troncheons ſtill in the pearth, for thei will
put foꝛth many motions, the whiche if thei ſhall haue ſcantie

F.ſ.   of

of roote, then dung your trouncheons within with good earth, and likewise aboue also, and thei shall doe well.

The tyme meete to cut Cions.

YE shall vnderstande that all Trees the whiche commonly doe put foorth Cions, if ye cut them in Winter, thei will put forth and spryng more aboundantly, for then thei bee all good to set and plant.

To set Bushe trees, as Gooseberies, or small Reisons.

THere be many other kinde of Bush trees, whiche will growe of Cions pricked in the grounde, as the Goose berie tree, the small Reison tree, the Barberie tree, the Blackthorne tree, these with many others, to bee planted in Winter, will growe without rootes : ye must also propne them and thei will take well enough: so likewise ye mate prick (in Marche) of Oziars in moyst groundes, and thei wil grow, and serue to many purposes for your Garden.

¶The fifth Chapter treateth of fower maner of Graffynges.

IT is to bee vnderstoode that there bee many waies of Graffynges, whereof I haue here onely put fower sortes, the whiche bee good, bothe sure and well proued, and easie to doe, the whiche ye maie vse well in two partes of the yere and more, for I haue (saieth he) graffed in our house, in euery Moneth, except October and Nouember, and thei haue taken well, whiche I haue (saieth he) in the Winter begunne to graffe, and in the Sommer graft in the Scutchine or Shielde accordyng to the tyme, forwarde or slowe: for certaine trees, specially yong faire Cions haue enough or more of their sappe vnto midd August, then others some had at Midsommer before.

The first maie to graffe all sortes of Trees.

AND first of all it is to be noted, that all sortes of franke Trees, as also wilde Trees of nature, maie bee graft with

with Graffes, and in the Scutchine, and bothe vds wall take, but specially those Trees whiche be of like nature: therefore it is better so to graffe: howbeit, thei maie well growe and take of other sorts of trees, but certaine trees be not so good, nor will prosper so well in the ende.

¶ *How to graffe Apple trees, Peare trees, Quince trees, and Medler trees.*

They graffe the Peare graffe, on other Peare stockes, and Apple, vpon Apple stocke, Crabbe or Vplsdyng stocke, the Quince and Medler, vppon the white Thorne, but most commonly thei vse to graffe one Apple vppon another, and bothe Peares and Quinces., thei graffe on Hawthorne and Crabbe stocke. An other kinde of fruite called in Frenche, *Saulsay*, thei vse to graffe on the Willowe stocke, the maner thereof is harde to doe, whiche I haue not seene, therefore I will let passe at this present.

The graffyng of greate Cheries.

They graffe the greate Cherie, called in Frenche *Heaulmiers*, vpon the Crabbe stocke, and an other long Cherie called *Guyniers*, vppon the wilde or sower Cherie tree, and likewise one Cherie vpon an other.

To graffe Medlers.

THE Misple or Medlar, thei maie bee graffed on other Medlars, or on white Thorne: the Quince is graffed on the white or blacke Thorne, and thei doe prosper well. I haue graffed (saieth he) the Quince vpon a wilde Peare stocke, and it hath taken and borne fruite well and good, but thei will not long endure. I beleue (saieth he) it was because that the graffe was not able enough to draw the sapp from the Peare stock. Some graffe the Medlar on the Quince, to bee greate. And it is to be noted, although the stocke and the graffe be of contrarie natures: yet notwithstandyng, neither the Graffe nor Scutchin, shall take any parte of the nature of the wilde stock so graffed, though it bee Peare, Apple, or Quince, whiche is contrarie againstt many whiche haue written, that if ye graffe the Medlar vpon the Quince tree, thei shalbe without stones,

F.ij.   whiche

whiche is abuſion and mockerie. For I haue (ſaieth he) proued the contrarie my ſelf.

Of diuers kindes of graffes.

IT is verie true, that one maie ſet a tree, whiche ſhall beare
diuers ſoꝛtes of fruite at once, if he be graffed with diuers
kinde of graffes, as the blacke, white, and greene Cherie
togethers, and alſo Apples of other Trees, as Apples and
Peares togethers, and in the Scutchiõ (ye maie graffe) likewiſe of diuers kindes alſo, as on Peares, Abꝛicotes, and
Plums together, and of others alſo.

Of the graffyng the Figge.

YE maie graffe the Figge tree vpon the Peache tree oꝛ A-
bꝛicote, but leaue a bꝛaunche on the ſtocke, and that muſt
bee accoꝛdyng foꝛ the ſpace of yeres, foꝛ the one ſhalt chaunge
ſoner then the other. All trees aboue ſaied, doe take very well
being graffed one with the other. And I haue not knowne, oꝛ
founde of any others, howbeeit (ſaieth he) I haue curiouſly
ſought and pꝛoued, becauſe thei ſaie one maie graffe on Cole-
woꝛtes, oꝛ on Elmes, the whiche I thinke are but ieſtes.

Of the greate Abricotes.

THe great Abꝛicote thei graffe in Sommer, in the Scutchion oꝛ ſhield, in the ſappe oꝛ barke of the leſſer Abꝛicote,
and be graffed on Peache trees, Figge trees, & principally on
Damſon oꝛ Plum trees, foꝛ there thei will pꝛoſper the better.

Of the Seruice tree.

OF the Seruice Tree, thei ſaie and wꝛite, that thei maie
hardly bee graft on other ſeruice Trees, either on Apple
trees, Peare, oꝛ Quince trees: and I belieue this to bee verie
hard to do, foꝛ I haue tried (ſaith he) and thei would not pꝛoue.

The ſettyng of Seruices.

THerefoꝛe it is muche better to ſet them of curnelles, as it
is afoꝛeſaied, as alſo in the ſeconde Chapter of the Plantyng of Cyons, oꝛ other greate trees, whiche muſt be cutte in
Winter, as ſuche as ſhalbe moſte meete foꝛ that purpoſe.

Trees whiche be verie harde to be graffed,
in the ſhield or ſcutchion.

All

ALL other maner of Trees aforefaied, doe take verie well to be graffed with Cions, and alfo in the fhield, except A-bricotes on Peaches, Almondes, Percigniers, the Peache tree doe take hardly to bee graffed, but in the fhield in Som-mer, as fhalbe more largely hereafter declared. As for the Al-monde, Percigniers and Peaches, ye maie better fet them of Curnelles and Nuttes, whereby thei fhall the foner come to perfection to be graffed.

How a man ought to confider thofe trees, whiche
be commonly charged with fruite.

WE fhall vnderftande, that in the beginnyng of graffyng, ye muft confider what fortes of trees, doe mofte charge the ftocke with braunche and fruite, or that doe loue the countrie or grounde, whereas you intende to plante or graffe them : for better it were to haue abundaunce of fruite, then to haue verie fewe or none good.

Of trees whereon to choofe your graffes.

OF fuch trees as ye wil gather your graffes to graffe with, ye muft take the at the endes of the principall braunches, whiche bee alfo faire and greateft of Sappe, hauing twoo or three fingers length of the old wood, with the newe, and thofe Cions whiche haue of eyes fomewhat nigh together, are the beft, for thofe whiche bee long, or farre one from an other, bee not fo good for to bryng fruite.

The Cions toward the Eaft are beft.

YE fhall vnderftande, that thofe Cions whiche do growe on the Eaft, or Orient part of the tree, are beft:ye muste not lightly gather of the euill and flender graffes, which growe in the middes of the trees, nor any graffes whiche doe growe within on the braunches , or that doe fpring from the ftocke of the tree,nor yet graffes whiche be on verie old trees, for thereby ye fhall not lightly profite to any purpofe.

To choofe your tree for Graffes.

AND when the trees,where as you intende to gather your graffes,be fmall and yong,as of fiue or fire yeres grouth, doe not take of the higheft graffe thereof, nor the greateft, excepte it bee of a small tree of twoo oꝛ thꝛee peres, the whiche commonly hath too muche of top oꝛ wood, otherwise not, foꝛ you shall but marre your graffyng.

How to keepe Graffes a long tyme.

YE maie keepe graffes a long tyme good, as from Alhallowtide (so that the leaues be falle) vnto the tyme of graffing, if that thei be well couered in the pearth halfe a foote depe therein, and so that none of the doe appere without the pearth.

How to keepe Graffes before thei are budded.

YE shall not gather them, excepte ye haue greate neede, vntill Chꝛistmas oꝛ there aboutes, and putte them not in the ground nigh any walles, foꝛ feare of Moles, Mice, and water marryng the place and graffes. It shalbe good to keepe graffes in the pearth befoꝛe thei beginne to bud, when that ye will graffe betwixte the barke and the tree, and when the trees beginne to enter into their sappe.

How one ought to begin to Graffe.

YE maie well beginne to graffe (in cleauyng the stocke) at Chꝛistmas, oꝛ befoꝛe, accoꝛdyng to the coldnesse of the tyme, and pꝛincipally the Pealme oꝛ greate Cherie, Peares, Wardens, oꝛ foꝛward fruite of Apples: and foꝛ Medlars it is good to tary vntill the ende of Januarie, and Febꝛuarie, vntill Marche, oꝛ vntill suche tyme as ye shall see the trees beginne to bud oꝛ spꝛing.

When it is good Graffyng the wilde stockes.

IN the Spꝛing tyme it is good graffyng of wilde stockes, (whiche be greate) betwixt the barke and the Tree, suche stockes as be of a lateward spꝛing, and kept in the pearth befoꝛe. The Damson oꝛ Plum taryeth longest to bee Grafte: foꝛ thei doe not shewe oꝛ put foꝛthe sap, so sone as the others.

Marke if the tree be forward or not.

YE ought to consider alwaies, whether the tree be foꝛward oꝛ not, oꝛ to bee graffed sone oꝛ lateward, and to giue hym also a graffe of the like hast oꝛ slownesse: euē so ye must marke the tyme, whether it be slowe oꝛ foꝛward.

When

When one will Graffe, what necessaries he
ought to be furnished withall.

VVHenſoeuer ye goe to graffyng, ſee ye be firſt furniſhed with graffes, claie and moſſe, clothes, oʒ barkes of ſal-lowe to binde the graffes, oʒ clouen Bʒiars, oʒ ſmall Oʒiers to binde likewiſe withall. Alſo ye muſt haue a ſmall Sawe, and a ſharpe knife, to cleaue and cutte graffes withall. But it were muche better if ye ſhould cut your graffes with a greate penknife, oʒ ſome other like ſharpe knife, hauyng alſo a ſmall wedge of hard wood, oʒ of Jron, with a hooked knife, and alſo a ſmall Mallet. And your wilde ſtockes muſt bee well rooted befoʒe ye doe graffe them: and be not ſo quicke to deceiue your ſelues, as thoſe whiche doe graffe and plant all at one tyme, yet thei ſhall not pʒofite ſo wel, foʒ where the wilde ſtocke hath not ſubſtaunce in hymſelf, muche leſſe to giue vnto the other graffes; foʒ when a man thinkes ſometymes to foʒward hym ſelf, he doeth hinder hymſelf.

Of graffes not proſperyng the firſt yere.

YE ſhall vnderſtande, that very hardly pour graffes ſhall pʒoſper after if thei doe not pʒofite oʒ pʒoſper well in the firſt pere, foʒ when ſoeuer (in the firſt pere) thei pʒofite well, it were better to graffe them ſomwhat lower then to let them ſo remaine and growe.

For to graffe well and ſounde.

AND foʒ the beſt vnderſtandyng of graffyng in the cleft, ye ſhall firſt cut awaie all the ſmall Cions about the bodie of the ſtocke beneath, and befoʒe ye beginne to cleaue pour ſtocke, dʒeſſe and cut your graffes ſomewhat thicke and readie, then cleaue your ſtocke, and as the cleft is ſmall oʒ greate (if neede be) pare it ſmooth within, then cut your inciſion of your graffes accoʒdyngly, and ſet them in the cleftes as euen and as cloſe as ye can poſſible.

How to trim your graffes.

YE maie graffe your Graffes full as long as twoo oʒ thʒee trunchions oʒ cut graffes, whiche ye maie like-wiſe graffe withall very well, and be as good as thoſe whiche

whiche doe come of olde wood, and oftentymes better, as to graffe a bough, for often it so happeneth, a man shall finde of Oylettes or eyes harde by the olde slender wood, yet better it were to cut them of with the olde wood, and choose a better and faire place at some other eye in the same graffe, and to make your incision there vnder, as aforesaied, and cut your graffes in makyng the incision on the one side narrowe, and on the other side brode, and the inner side thinne, and the cut side thicke, because the outside (of your graffe) must ioyne within the cleft, with the sappe or barke of the wilde stocke, and it shall so bee set in: see also that ye cut it smooth as your cleftes are in the stocke, in ioynyng at euery place bothe euen and close, and especially the ioyntes or corners of the graffes on the head of the stocke, whiche must be well and cleane pared before, and then set fast thereon.

How to cut graffes for Cheries and Plummes.

It is not muche requisite in the Healme Cherie, for to ioyne the graffes (in the stocke) wholy throughout, as it is in others, or to cut the graffes of greate Cheries, Damsons or Plummes, so thinne and plaine as ye make other graffes, for these sorte haue a more greater sappe or pithe within, the whiche ye must alwaies take heede in cuttyng it to nigh on the one side, or on the other, but at the ende thereof chiefly, to be thinne cut and flatt.

Note also.

And yet if the saied incision bee more straighter and closer on the one side then on the other side, parte it where it is most meete, and where it is to straight, open it with a wedge of Iron, and put in a wedge of the same wood aboue in the cleft, and thus make ye moderate your graffes as ye shall see cause.

How in graffyng to take heede that the barke doe not rise

IN all kinde of cutting your graffes, take heede to the barke of your graffes, that it doe not rise (from the wood) on no side thereof, and specially on the outside, therfore ye shall leaue

it

it moꝛe thicker then the inner ſide : Alſo ye muſt take heede when as the ſtocks doe wꝛeath in cleauing, that ye may ioyne the graffe therein accoꝛdyngly : the beſt remedie therefoꝛe is to cut it ſmooth within, that the graffe maie ioyne the better : ye ſhall alſo vnto the moſt greateſt ſtockes, chooſe foꝛ them the moſt greateſt graffes.

How to cut your ſtocke.

HOw muche the moꝛe your ſtocke is thinne and ſlender, ſo muche moꝛe ye ought to cut hym lower, and if your ſtocke bee as greate as your finger, oꝛ there aboutes, ye maie cut hym a foote oꝛ half a foote from the earth, and digge hym about, and dung hym with Goates dung, to helpe hym withall, and graffe hym but with one Graffe oꝛ Cion.

If the wilde ſtocke be greate and ſlender.

IF your wilde ſtocke be great, oꝛ as bigge as a good ſtaffe, ye ſhall cut hym rounde of, a foote oꝛ there aboutes aboue the yearth, then ſet in two good graffes in the head oꝛ cleft thereof.

Trees as greate as ones arme.

AND when your ſtocke is as greate as your arme, ye ſhall ſawe hym cleane of round, thꝛee oꝛ fower foote, oꝛ there aboutes from the yearth, foꝛ to defende hym, and ſet in the head thꝛee graffes, twoo in the cleft, and one betwixt the Barke and the Tree, on that ſide whiche ye maie haue moſt ſpace.

Greate Trees as bigge as your legge.

IF the ſtocke bee as bigge as your legge, oꝛ there aboutes, ye ſhall ſawe hym faire and cleane of, fower oꝛ fiue foote hie from the yearth, and cleaue hym a croſſe (if ye will) and ſet in fower graffes in the cleftes thereof, oꝛ els one cleft onely, and ſet two graffes in bothe the ſides thereof, and other two graffes betwixt the barke and the Tree.

When the Graffes be pinched with the Stocke.

G. j.   Ye

Ye muſt foz the better vnderſtandyng, marke to graffe betwixt the barke and the tree, foz when the ſappe is full in the wood of wilde ſtockes beyng great, then thei doe commonly pinche oz wzing the graffes to ſoze, if ye doe not put a ſmall wedge of greene wood in the cleft thereof, to helpe them withall againſt ſuche daunger.

How ye ought to cleaue your ſtockes.

Whẽ ſoeuer ye ſhall cleaue your wilde ſtockes, take heede that ye cleaue them not in the middeſt of the harte oz pithe, but a little on the one ſide, whiche ye ſhall thinke good.

cut not the ſto-
ke in ẙ middeſt

How to graffe the braunche of greate Trees.

Whẽ ſoeuer ye would graffe great trees, as great as your thigh, oz greater, it were muche better to graffe onely the bzaunches thereof, then the ſtocke oz bodie, foz the ſtocke will rotte befoze the graffes ſhall couer the head.

How to cut braunches olde and greate.

But if the bzaunches be to rude, and without ozder (the beſt ſhall be) to cut them all of, and within thzee oz foure yeres after thei will bzing faire newe Cions againe, and then it ſhall bee beſt to graffe them, and cut of all the ſuperfluous and ill bzaunches thereof.

¶ How ye ought to binde your graffes throughout for feare of windes.

And when your graffes ſhall be growne, ye muſt binde them, foz feare of ſhakyng of the winde, and if the tree be free and good of hym ſelf, let the Cions grow ſtill, and ye maie graffe any parte oz bzaunche ye will in the cleft, oz betwixt the barke and the tree, either in the Scutchion, and if your barke be faire and looſe.

To ſet many graffes in one cleſt.

When ye will put many graffes in one cleſt, ſee that one inciſion (of your graffe) bee as large as the other, not to be put into the cleft ſo ſlightly and raſhly, and that one ſide thereof be not moze open then the other,

and

and that thefe graffes be all of one length: it ſhall ſuffice alſo,
if thei haue three eyes on eche graffe without the ioynt therof.

How to ſawe your ſtocke, before ye leaue hym.

IN ſawyng your ſtocke, ſee that ye teare not the barke a-
bout the head thereof, then cleaue his head with a long
ſharpe knife, or ſuche like, and knocke your wedge in the
mioueſt thereof, (then pare him on the head rounde about) and
knocke your wedge in ſo deepe till it open meete for your
Graffes, but not ſo wide, then holdyng in one hande your
graffe, and in the other hande your ſtocke, ſett your graffe in
cloſe, barke to barke, and let your wedge bee greate aboue at
the head, that ye maie knocke hym out faire and caſly againe.

*⁊ If the ſtocke cleaue too muche, or
the barke doe open.*

IF the ſtocke doe cleaue too muche, or open the barke with
the wood too lowe, then ſoftly open your ſtocke with your
wedge, and ſee if your inciſion of your graffe, be all meete and
iuſte, accordyng to the cleft, if not, make it vntill it be meete, or
els ſawe hym of lower.

⁊ How graffes neuer lightly take.

ABoue all thinges, ye muſt conſider the metyng of the two
ſappes, betwixte the graffe and the wilde ſtocke, whiche
muſt be ſet iuſte one with an other: for ye ſhall vnderſtande, if
thei doe not ioyne, and the one delight with the other, beeyng
euen ſette, thei ſhall neuer take together, for there is nothyng
onely to ioyne their increaſe, but the Sappe, recountyng the
one againſt the other.

⁊ How to ſet the graffes right in the cleft.

WHen the barke of the ſtocke, is more thicker then the
graffe, ye muſt take good heede, of the ſettyng in of the
graffe in the clefte, to the ende that his ſappe maie ioyne right
with the ſappe of the ſtocke, on the in ſide, and ye ought likewiſe to conſider of the ſappe of the ſtocke, if he doe ſurmounte
the graffes on the out ſides of the cleft too muche or not.

⁊ Of ſettyng in the graffes.

G.ij.　　　Alſo

ALſo ye muſt take good heede, that the graffes be well and cleane ſet in, and ioyne cloſe vpon the hedde of the ſtocke: likewiſe then the inciſion whiche is ſet in the clefte, doe ioyne very well within on bothe ſides, not to ioyne ſo euen, but ſomtymes it maie doe ſeruice, when as the graffes doe drawe too muche from the ſtocke, or the ſtocke alſo on the Graffes doe put forthe.

¶ Note alſo.

ANd therefore, when the ſtocke is rightly clouen, there is no daunger in cuttyng the inciſion of the graffe, but a little ſtraight rebated to the ende thereof, that the ſappe maie ioyne one with the other, the better and cloſer together.

¶ How ye ought to drawe out your wedge.

VVHen your graffes ſhalbe well ioyned within the ſtock, drawe your wedge faire and ſoftly forthe, for feare of diſplacyng your graffes, ye maie leaue within the cleft a ſmal wedge of ſuche greene wood, as is aforeſaied, and ye ſhall cut it of cloſe by the hedde of your ſtocke, and ſo couer it with a barke as followeth.

¶ To couer your cleftes on the hedde.

VVHen your wedge is drawen forthe, put a greene pill of thicke barke of Willowe, Crabbe, or Apple, vppon your cleftes of the ſtocke, that nothyng maie fall betweene: then couer all aboute the cleftes on the ſtocke hedde, twoo fingers thicke with good claie, or nigh about that thickneſſe, that no Winde nor Raine maie enter. Then couer it rounde with good Moſſe, and then wreathe it ouer with clothes, or pilles of Willowe, Brier, of Oziars, or ſuche like, then binde them faſte, and ſticke certaine long prickes on the Graftes hedde emongeſt your Cions, to defende them from the Crowes, Iayes, or ſuche like.

¶ How ye ought to ſee to the bindyng of your graffes.

BUt alwaies take good heede to the bindyng of your heads that thei ware ſlacke, or ſhagge, neither on the one ſide or other, but remaine faſt vpon the claye, whiche claye remaines faſt

faſt(likewiſe on the ſtocke hedde)vnder the bindyng thereof, wherefoꝛe,the ſaide claye muſt be moderated in ſuche ſoꝛte as followeth.

¶ How you ought to temper your claye.

THe beſt waie is therefoꝛe,to trie your claye betwixt your handes,foꝛ ſtones and ſuche like,and ſo to temper it as ye ſhall thinke good, if ſo it require of moiſtneſſe oꝛ dꝛineſſe,and to temper it with the haire of beaſtes:foꝛ when it dꝛieth,it holdeth not (otherwiſe) ſo well on the ſtocke,oꝛ if ye kneade of Moſſe therewith,oꝛ mingle Hate thinne therewith: ſome doe iudge, that the Moſſe doeth make the Trees moſſie . But J thinke(ſaieth he)that commeth of the diſpoſition of places.

To buſhe your graffe heddes.

WHen ye ſhall binde oꝛ wꝛappe your Graffe heddes with bande , take ſmall Thoꝛnes , and binde them within,foꝛ to defende your Graffes from Kites,oꝛ Crowes,oꝛ other daunger of other foules,oꝛ pꝛicke of ſharpe white ſtickes thereon.

The ſeconde waie to graffe high braunches on Trees.

THE ſeconde maner to Graffe,is ſtraunge inough to many : This kinde of Graffyng is on the toppes of bꝛaunchs of Trees , whiche thyng to make theim growe lightly,is not ſone obtained: whereſoeuer thei be graffed , thei doe onely require a faire yong wood , a greate Cion oꝛ twigge,growing higheſt in the Tree toppe,whiche Cions ye ſhall chooſe to graſſe on,of many ſoꝛtes of fruites if ye will oꝛ as ye ſhall thinke good,whiche oꝛder followeth.

TAke Graffes of other ſoꝛtes of Trees , whiche ye would graffe in the topp thereof,then mount to the toppe of the Tree whiche ye would Graffe , and cut of ẏ tops of all ſuch bꝛaunches,oꝛ as many as ye would Graffe on,and if thei bee greater then the Graffes,whiche ye would Graffe , ye ſhall cutte and Graffe them lower , as ye doe the ſmall wilde ſtocke afoꝛeſaid.But if the Cions that you cutte,

bee as greate as as your Graffe that you Graffe on, ye shall
cutte them lower betwixt the old wood and the newe, or a little more higher, or lower: then cleaue a little, and choose your
Graffes in the like sorte, whiche ye would Plant, whereof ye
shall make the incision shorte, with the barke on bothe sides
like, and as thicke on the one side as the other, and sett so iuste
in the clefte, that the barke maie bee euen and close, as well a-
boue as beneath, on the one side as the other, and so binde hym
as is aforesaied. It shall suffice that euery Graffe haue an oylet, or eye, or twoo at the moste, without the ioynt, for to
leaue them too long it shall not bee good, and ye must dresse
it with Claye and Mosse, and binde it, as is aforesaied. And
likewise ye maie Graffe these, as ye doe the little wilde stockes, whiche should bee as greate as your Graffes, and to
Graffe them, as ye doe those with Sappe like on bothe sides,
but then ye must graffe them in the yearth, as three fingers
of, or there aboutes.

*The maner of Graffyng, is of Graffes whiche
maie bee sette betwixt the barke
and the Tree.*

*To graffe betwixt the barke
and the tree.*

This maner of Graffyng is good, when Trees
doe beginne to enter into their Sappe, whiche
is, aboute the ende of Februarie, vnto the ende
of Aprill, and specially on greate wilde stockes,
whiche bee hard to cleaue, ye maie set in fower
or fiue Graffes in the hedde thereof, whiche Graffes ought to
bee gathered afore, and kepte close in the yearth till then, for
by that tyme aforesaied, ye shall scantly finde a Tree, but that
he doeth put forthe or budde, as the Apple called *Capendu*, or
suche like. Ye must therefore sawe these wilde stockes more
charily, and more higher, so thei bee greate, and then cutte the

Graffes

Graffes, whiche ye would set together, so as you would sette them vpon the wilde stocke that is clefte, as is afore rehearsed. And the incision of your Graffes must not bee so long, nor so thicke, and the barke a little at the ende thereof must bee taken awaie, and made in maner as a Launcet of Jron, and as thicke on the one side as the other.

How to dresse the head, to place the graffes betwixt the barke and the tree.

AND when your Graffes bee readie cut, then shall ye cleanse the hed of your stock, and pare it with a sharpe knife, rounde about the barke thereof, to the ende your graffes maie ioyne the better thereon, then by and by take a sharpe penknife, or other sharpe poynted knife, and thrust it doune betwixt the barke and the stocke, so long as the incision of your graffes be, then put your graffes softly doune therein to the hard ioynt : and see that it doe sit close vppon the stocke head.

How to couer the head of your stocke.

WHen as ye haue set in your graffes, ye must then couer it wel about with good tough Claie and Mosse, as is saied of the others, and then ye must incontinent enuyron or compasse your head with small thornie bushes, and binde them fast thereon all about, for feare of greate birdes, and likewise the winde.

¶ Of the maner and graffyng in the Shielde or Scutchion.

THE fourth maner to graffe, whiche is the last, is to graffe in the Scutchion, in the sappe, in Sommer, from about the ende of the Moneth of Maie, vntill August, when as Trees be yet strong in sappe and leaues, for other waies it can not bee doen, the best tyme is in June and July, so it is some yeares when the tyme is very drie, that some Trees doe holde their sappe very long, therfore ye must tarie till it returne.

For to graffe in Sommer so long as the Trees be fullleaued.

.F oz

FOR to beginne this maner of graffyng well, ye must in Sommer when the Trees bee almost full of sappe, and when thei haue sprong forth of newe shootes being somewhat hardned, then shall ye take a braunche thereof in the toppe of the Tree, the whiche ye will haue graffed, and choose the highest and the principallest braunches, without cuttyng it from the old wood, and choose thereof the principallest oylet or eye, or buddyng place, of eche braunche one, with whiche oylet or eye, ye shall beginne to graffe, as followeth.

The bigge Cions are best to graffe.

PRincipally ye must vnderstande, that the smallest and naughtie oplettes or buddes of the saied Cions, be not so good to graffe, therefore choose the greatest and best ye can finde, first cut of the leafe harde by the oylet, then ye shall trenche or cut (the length of a Barley corne) beneath the oylet rounde about the barke, harde to the wood, and so like-wise aboue: then with a sharpe point of a knife, slit it doune halfe an ynche beside the oylet or budde, and with the poynt of a sharpe knife softly raise the saied shielde or Scutchion round about, with the oylet in the middest, and all the sappe belongyng thereunto.

How to take of the Shielde from the wood.

AND for the better raising the saied shielde or Scut-chion from the wood, after that ye haue cut him rond about, and then slit hym doune, without cuttyng any parte of the wood within, ye must then raise the side next you that is slit, and then take the same shielde betwixt your finger and thumbe, and plucke or raise it softly of, without brea-kyng or brusing any parse thereof, and in the opening or pluc-kyng it of, holde it with your finger harde to the wood, to the ende the sappe of the oylet maie remaine in the shielde, for if it goe of (in pluckyng it) from the barkes, and sticke to the wood, your Scutchion is nothyng worth.

¶ To knowe your Scutchion or shielde, when he is good or badde.

And

ND for the moze easier vnderstanding, if it be good
oz badde, when it is taken from the wood, looke with
in the saied shielde, and if ye shall see it cracke, oz o·
pen within, then it is of no value; foz the chiefe sappe doeth yet
remaine behinde with the wood, which should be in the shield,
and therefoze ye must choose and cut an other shielde, whiche
must be good and sounde, as afozesaied, and when your Scut=
chion shall be well taken of from the wood, then holde it dzie
by the oylet oz eye betwixt your lippes, vntill ye haue cut and
taken of the barke from the other Cion oz bzaunche, and set
hym in that place, and looke that ye doe not foule oz wet it in
your mouth.

Of yong Trees to graffe on.

UT ye must graffe on suche Trees, as bee from the
bignesse of your little finger, vnto as greate as your
arme, hauyng their barke thinne and slender, foz
greate Trees commonly haue their barke harde and thicke,
which ye can not well graffe this waie, except thei haue some
bzaunches with a thinne smooth barke, meete foz this waie
to be doen.

How to set or place your shielde.

YE must quickly cut of round the barke of the tree that
ye will graffe on, a little moze longer then the shielde
that ye set on, because it maie ioyne the sooner and ea=
sier, but take heede that in cuttyng of the barke, ye cut not the
wood within.

Note also

Fter the incision once doen, ye must then couer bothe
the sides oz endes wel and softly withall, with a little
bone oz hozne, made in maner like a thinne skinne,
whiche ye shall laie it all ouer the ioyntes oz closinges of the
saied shielde, somewhat longer and larger, but take heede foz
hurtyng oz crushyng the barke thereof.

How to lift vp the barke, and to set your Shielde on.

H.j.   This

This doen, take your shielde oʒ Scutchion, by the oylet oʒ eye that he hath, and open him faire and softly by the twoo sides, and put them straight waie on the other tree, whereas the barke is taken of, and ioyne him close barke to barke thereon, then plaine it softly aboue and at bothe the endes with the thinne bone, and that thei ioyne aboue and beneath barke to barke, so that he maie feede well the bʒaunche of that tree.

How to binde on your shielde.

This doen, ye must haue a wʒeathe of good Hempe, to binde the saied shielde on his place : the maner to binde it is this, ye shall make a wʒeath of Hempe together as greate as a Goose quill, oʒ there aboutes, oʒ accoʒdyng to the bignesse oʒ smalnesse of your tree: then take your Hempe in the mioff, that the one halfe maie serue foʒ the vpper halfe of the shieloe, in windyng and crossyng (with the Hempe) the saied shielde, on the bʒaunch of the Tree, but let that ye binde it not to straight, foʒ it shall let hym from taking oʒ spʒinyng, and likewise their sappe cannot easily come oʒ passe from the one to the other: and see also that wet come not to your shield, noʒ likewise the Hempe that ye binde it withall. Ye shall beginne to binde your Scutchion firff behinde in the middeff of your shielde, in commyng still lower and lower, and so recouer vnder the oylet and taile of your shielde, bindyng it nigh togethers, without recoueryng of the saied oylet, then ye shall returne againe vpwarde, in bindyng it backwarde to the middeff where ye began. Then take the other part of the Hempe, and binde so likewise the vpper parte of your shielde, and encreafe your Hempe as ye shall neede, and so returne againe backwarde, and ye shall binde it so, till the fruites oʒ cleftes be couered (bothe aboue and beneath) with your saied Hempe, except the oylet and his taile, the whiche ye muff not couer, foʒ that taile will shed aparte, if the shielde doe take.

On one Tree ye maie graffe or put two or three shieldes.

Ye

€ maie verp well if pe will, on euerp tree graffe two
o; thꝛee ſhieldes, but ſee that one be not right againſt
an other, noꝛ pet of the one ſide of the Tree, let pour
ſhieldes ſo remaine bounde on the trees, one Moneth oꝛ moꝛe
after thei be graffed, and the greater the Tree is, the longer
to remaine, and the ſmaller the leſſer tyme.

The tyme to vnbinde your Shield.

AND then after one Moneth, oꝛ ſixe Weekes paſte,
pe muſt vnbinde the Shield, oꝛ at the leaſt, cutte the
hempe behinde the Tree, and let it ſo remaine vnto
the Winter nert followyng, and then aboute the Monethe of
Marche, oꝛ Apꝛill if pe will, oꝛ when pe ſhall ſee the Sappe of
the Shield put foꝛthe, then cut the bꝛanche aboue the Shield,
thꝛee fingers all about all of.

How to cutte and gouerne the braun-
ches, graffed on the Trees.

THen in the nerte pere after that the Cions ſhall bee
well ſtrengthened, and when thei doe begin to ſpꝛing,
then ſhall pe cutte them all harde of, by the Shield a-
boue, foꝛ if pe had cutte them ſo nigh in the firſte pere, when
thei begamne firſt to ſpꝛyng oꝛ budde, it ſhould greatly hinder
them, againſt their increaſe of growyng: alſo when thoſe Cions ſhall put foꝛthe a faire wood, pe muſt binde and ſtaie them
in the middes, faire and gently with ſmall wandes, oꝛ ſuche
like, that the Winde and weather hurte them not. And after
this maner of Graffyng, is pꝛactiſed in the Shield oꝛ Scutchion, whiche waie pe maie eaſily Graffe the white Roſe on
the redde: and likewiſe pe maie haue Roſes of diuers colours
and ſoꝛtes, vpon one bꝛaunche oꝛ roote. This I thought ſufficient and meete to declare, of this kinde of Graffyng at this
pꝛeſent.

¶ The ſixt Chapter is of tranſplantyng
or alteryng of Trees.

H.ij.   The

The soner ye transplant or set them,
it shall bee the better.

E ought to tranſplant or ſet your trees, from Alhallowtide vnto Marche, and the ſoner the better, for as ſone as the leaues are fallen from the Trees, thei bee meete for to bee Planted, if it bee not in a verie colde or moiſte place, the whiche then it were beſte for to tarie vnto Ianuarie, or Februarie: to Plant in the Froſt is not good.

To plant or ſet towardes the Southe,
or Sunnie place is beſt.

Afore ye doe plucke vp your Trees for to plant them, if ye will marke the Southſide of eche Tree, that when ye ſhall replant them, ye maie ſet them againe as thei ſtoode before, whiche is the beſt waie as ſome doe ſaie. Alſo if ye keepe them a certaine tyme, after thei be taken out of the yearth, before ye replante theim againe, thei will rather recouer there in the yearth, ſo thei bee not wette with raine, nor otherwiſe, for that ſhall bee more contrary to theim, then the greate heate or drought.

How to cutte the braunches of Trees,
before thei bee ſet.

Whenſoeuer ye ſhall ſette, or replant your Trees, firſt ye muſte cutte of the boughes, and ſpecially thoſe whiche are greate braunches, in ſuche ſorte, that ye ſhall leaue the ſmall twigges or ſprigges, on the ſtockes of your braunche, whiche muſt bee but a ſhaftment long, or ſomwhat more or leſſe, accordyng as the Tree ſhall require, whiche ye doe ſette.

Apple Trees commonly muſt be diſbraunched
before thei bee replanted or ſette.

And chieflly the Apple Trees, beyng Graffed or not Graffed, doe require to bee diſbraunched before thei bee ſette againe, for thei ſhall proſper thereby, muche
the

the better : the other sortes of Trees maie well passe vnbraun=
cheo,if thei haue not too greate oz large bzaunches; ano therefoze,it shall be good to transplant oz sette,as soue after as the
Graffes are closeo , on the hedde of the wilde stocke , as foz
small Trees,whiche haue but one Cion oz twigge, it needes
not to cutte them aboue,when thei bee replanteo oz remoueo.

All wilde stockes must bee disbraunched,
when thei are replanted or sette.

ALL wilde Trees oz stockes, whiche ye thinke foz
to Graffe on,ye must first cutte of all their bzaunches befoze ye sette theim againe : also it shall bee
good, alwaies to take heede in replantyng your
Trees, that ye doe sette theim againe, in as good oz better
earth, then thei were in befoze,ano so euery Tree, accozdyng
as his nature doeth require.

¶ *What Trees loue the faire Sunne,*
what Trees the cold aire.

COmmonly the moste parte of Trees, doe loue the
Sunne at Noone, ano yet the Southe winde (or
vent d'aual)is very contrary againi their nature,
ano specially the Almonde tree , the Abzicote , the
Mulberie tree , the Figge tree , and the Pomgranade Tree:
Certaine other trees there be,whiche loue cold aire,as these:
The Chestnut Tree, the wilde and eager Cherie Tree , the
Quince Tree,and the Damson oz Plum tree: The Walnut
loueth colde aire,and a stonie white grounde : Peare Trees
loue not greatly plaine places, thei prospere well inough in
places closeo with Walles,oz high Hedges,ano specially the
Peare called *bon Crestien.*

Of many sortes and maners of Trees,
followyng their nature.

THE Damson oz Plum Tree, doeth loue a cold
fatte yearth,ano claie withall,the(Healme)great
Cherie doeth loue tobee sette oz Planteo vppon
claie. The Pine Tree loueth light pearth, stonie
and Sandie . The Medlar commeth well inough in all kinde

H.iij. of

of groundes, and doeth not hinder his fruite, to bee in the sha¬
dowe and moiste places. Hasell nutte Trees loue the place to
bee cold, leane, moiste and Sandie. Ye shall vnderstande, that
euery kinde of fruitfull Tree doeth loue, and is more fruite¬
full in one place, then an other, as accordyng vnto their nature.
neuerthelesse, yet we ought to nourishe the (all that we maie)
in the place where wee sette them in, in takyng them fro the
place and ground thei were in. And ye must also consider whe
one doeth plant them, of the greate and largest kinde of trees,
that euery kinde of tree maie prospere and growe, and it is to
be considered also, if the Trees haue commonly growne afore
so large in that grounde or not, for in good yearth, the Trees
maie well prosper and growe, hauyng a good space one from
an other, more then if the grounde were leane and naught.

How to place or set Trees at large.

IN this thing ye shall consider, ye must giue a com¬
petent space, from one Tree to an other, when as
ye make the holes to set them in, not nigh, nor that
one Tree touche an other. For a good Tree Plan¬
ted, or set well at large, it profiteth oftentymes more of fruite
then three or fower Trees, sett too nigh together. The moste
greatest and largest Trees commonly are Walnuttes, and
Chestnuttes, if ye plante them seuerally in ranke, as thei doe
commonly growe vpon high waies, besides hedges and feel¬
des, thei must bee set xxxv. foote a sonder, one from an other, or
there aboutes, but if ye will plant many rankes in one place
togethers, ye must set them the space of xlv. foote, one from an
other, or there aboutes, and so farre ye must sette your rankes
one from an other. For the Peare Trees and Apple Trees,
and other sortes of Trees, whiche maie bee sette of this large¬
nesse one from the other, if ye doe Plante onely in rankes by
hedges in the feeldes, or otherwise, it shall bee sufficient of xx.
foote one from an other. But if ye will set twoo rankes vpon
the sides of your greate Alleis in Gardens, whiche bee of ten
or twelue foote broad, it shalbe then beste to giue them more
space, the one from the other in eche ranke, as about xxv. foote
also

alſo ye muſt not ſette your Trees right one againſt the other
but entermedlyng oʒ betweene euery ſpace, as thei maie beſt
growe at large, that if neede bee, ye maie plant of other ſmaller Trees betwene, but ſee that ye ſette them not to thicke. If
ye liſte to ſette oʒ plant all your Trees of one bigneſſe, as of
yong Trees like roddes, beyng Peare trees, oʒ Apple trees,
thei muſt be ſette a good ſpace one from an other, as of twentie oʒ thirtie foote in ſquare, as to ſaie, from one ranke to an o-
ther. Foʒ to plant oʒ ſette of ſmaller Trees, as Plum Trees,
and Apple Trees, of the like bigneſſe, it ſhalbee ſufficient foʒ
them fourteene oʒ fifteene foote ſpace, in quarters. But if ye
will plant oʒ ſette twoo rankes, in your Allepes in Gardens,
ye muſt deuiſe foʒ to proportion it after the largeneſſe of your
ſaid alleyes. Foʒ to plant oʒ ſette eager oʒ ſower Cherie trees
this ſpace ſhalbe ſufficient inough the one from the other, that
is, of tenne oʒ twelue foote, and therefoʒe if you make of great
oʒ large Allepes in your Garden, as of tenne foote wide, oʒ
there aboutes, thei ſhall come well to paſſe, and ſhall bee ſufficient to plant your Trees, of nine oʒ tenne foote ſpace: and foʒ
the other leſſer ſoʒtes of Trees, as of Quince Trees, Figge
Trees, Nutte Trees, and ſuche like, whiche bee not commonly planted, but in one ranke together.

Orderyng your Trees.

When that ye plant oʒ ſet rankes, of euery kinde of
Trees togethers, ye ſhall ſett oʒ plant the moſte
ſmalleſt towardes the Sunne, & the greateſt in the
ſhade, that thei maie not annoie oʒ hurte ẙ ſmall, noʒ the ſmall
the great. Alſo whēſoeuer ye will plant oʒ ſet of Peare trees,
and Plum trees (in any place) the one with an other, better it
were to ſett the Plum trees next the Sunne, foʒ the Peares
will dure better in the ſhade. Alſo ye muſt vnderſtande, when
ye ſet oʒ plant any rankes of Trees togethers, ye muſt haue
moʒe ſpace betwixt your rankes and trees, (then when ye ſet
but one ranke) that thei maie haue roome ſufficient on euery
ſide: Ye ſhall alſo ſcarſly ſett oʒ plant Peare trees, oʒ Apple
trees, oʒ other greate Trees, vppon dead oʒ moſſie barren
grounde

grounde buſtirred, foz thei increaſe (thereon) to no purpoſe.
But other leſſer trees verp well maie grow, as Plum trees,
and ſuche like: now when all the ſaied thynges aboue be confi=
dered; pe ſhall make pour holes accozdyng to the ſpace that
ſhall be required of euerp Tree that pe ſhall plant oz ſett, and
alſo the place meete foz the ſame ſo muche as pe maie conue=
nieht, pe ſhall make pour holes large enough, foz pe muſt ſup=
poſe the Tree pe doe ſett, hath not the halfe of his rootes he
ſhall haue hereaſter; therefoze pe muſt helpe him and giue him
of good fatt pearth, (oz dung) all about the rootes when as pe
plant him. And if anp of the ſame rootes be to long, and bzuſed
oz hurte, pe ſhall cut them cleane of a ſlope wiſe, ſo that the vp=
per ſide (of eche roote) ſo cut; maie be longeſt in ſettyng, and
foz the ſmall rootes whiche come foztje all about thereof, pe
maie not cut them of as the greate rootes.

*How ye ought to enlarge the holes for your
Trees, when ye plant them.*

VVhen as pe ſet the Trees in the holes, pe muſt then en=
large the rootes in placing them, and ſee that thei take
all dounewardes, without turnpng any rootes the ende vp=
ward, and pe muſt not plant oz ſet them to deepe in the earth,
but as pe ſhall ſee cauſe. It ſhall be ſufficient foz them to bee
planted oz ſet (halfe a foote, oz there aboutes) in the pearth, ſo
that the pearth bee aboue all the rootes halfe a foote oz moze,
if the place be not verp burnyng and ſtonie.

*Of dung and good yearth, for your
Plantes and Trees.*

AND when as pe would replant oz ſet, pe muſt haue of
good fatt pearth oz dung, well medled with a parte of
the ſame pearth whereas pe tooke pour Plantes out
of, with all the vpper creſtes of the pearth, as thicke as pe can
haue it: the ſaied pearth whiche pe ſhall put about the rootes,
muſt not bee put to nigh the rootes, foz doubt of the dung be=
png laied to nigh, whiche will put the ſaied rootes in a heate,
but let it be well medled with the other pearth, and welftrim=
pered in the holes, and the ſmalleſt and ſtewzeſt Cions that
turnes

turnes vp among those rootes, ye maie plant therewith very
well.

If ye haue wormes emongeſt the yearth
of your rootes.

IF there be wormes in the fat pearth oʒ dung, that ye put
about your rootes, ye muſt meddle it well alſo with the
dung of Oʒen oʒ Kine, oʒ ſiekt Sope aſhes aboute the
roote, which will make the wormes to dye, foʒ otherwiſe, thei
will hurt greatly the rootes.

¶ *To digge well the yearth about the*
Tree rootes.

ALſo ye muſt digge well the pearth, principally all
rounde ouer the rootes, and moʒe oftner if thei be
dʒie, then if thei bee wet, ye muſt not plant, oʒ ſett
Trees when it raineth, noʒ the pearth to bee very
moyſt about the rootes. The Trees that bee planted oʒ ſet in
vallies, commonly pʒoſper well by dʒougth, and when it raineth, thei that bee on the hilles are better by wateryng with
dʒoppes, then others, but if the place oʒ grounde be moyſt of
nature, ye muſt plant oʒ ſet your Trees ſo deepe thereon.

The nature of places.

ON high and dʒie places, ye muſt plant oʒ ſet your Trees
a little moʒe deeper, then in the vallies, and ye muſt not
fill the holes in high places, ſo full as the other, to the ende
that the raine maie better moyſten them.

Of good yearth.

YE ſhall vnderſtande that of good pearth, commonly
commeth good fruite, but in certaine places (if that
thei might be ſuffered to groWe) ther would ſeaſon the
Tree the better. Otherwiſe thei ſhall not come to pʒoofe, noʒ
yet haue a good taſte.

With what ye ought to binde your Trees.

VVhenſoeuer your Trees ſhall bee replanted oʒ ſett, ye
muſt knocke in (by the roote) a ſtake, and binde your
trees thereto foʒ ſeare of the winde: and when thei doe ſpring,
ye ſhall dʒeſſe them and binde them with bands that maie not

J.j.   bʒeake

bꝛeake, whiche bandes maie be of ſtrong ſoft hearbe, as Bul-
ruſhes oꝛ ſuche like, oꝛ of olde linnen cloutes, if the other bee
not ſtrong enough, oꝛ els ye maie binde them with Oziars, oꝛ
ſuche like, but foꝛ feare of frettyng oꝛ hurtyng your Trees.

¶ The vij. Chapter is of medecinyng and
keepyng the Trees when thei are planted.

¶ The firſt councell is, when your Trees bee but
Plantes (in drie weather) thei
muſt be watered.

THE yong Trees which be newly planted, muſt ſometymes (in Sommer) bee
watered when the tyme waxeth dꝛie, at
the leaſt the firſt yere after thei bee planted oꝛ ſet. But as foꝛ other greater trees
which are well taken and rooted a good
tyme, ye muſt digge them all ouer the
rootes after Alhaliowtide, and vncouer them ſower oꝛ fiue
foote compaſſe about the roote oꝛ tree: and let them ſo lye vncouered vntill the latter ende of Winter. And if ye doe then
meddle about eche Tree of good fatt yearth oꝛ dung, to heate
and comfoꝛte the yearth withall, it ſhall be good.

With what dung ye ought to dung your Trees.

A ND pꝛincipally vnto Moſſie Trees, dung them with
Hogs dung medled with other earth of the ſame ground,
and the dung of Oxen bee next about the rootes, and ye ſhall
alſo abate the Moſſe of the Trees with a great knife of wood,
oꝛ ſuche like, ſo that ye hurte not the barke thereof.

When ye ought to vncouer your
Trees in Sommer.

I N the tyme of Sommer, when the yearth is ſcantly halfe
moiſte, it ſhall be good to digge at the foote of the Trees,
all about on the roote, ſuche as haue not been vncouered in
the Winter befoꝛe, and to meddle it with good fatte yearth:
and

and so fill it againe, and thei shall doe well.

*When ye ought to cutte or
proine your Trees.*

AND if there bee in your Trees certaine braunches of superfluous wood, that ye will cutte of, tary vntill the tyme of the entryng in of the Sappe, that is, when thei beginne to budde, as in Marche and Aprill: Then cutte of as ye shall see cause, all suche superfluous braunches hard by the Tree, that thereby the other braunches maie prospere the better, for then thei shall soner close their sappe vpon the cutte places then in the Winter, whiche should not doe so well to cutte theim, as certaine doe teache, whiche haue not good experience. But for so muche as in this tyme the Trees bee entryng into the sappe, as is aforesaid. Take heede therefore in cuttyng then of your greate braunches hastely, that through their greate waight, thei doe not cleaue or separate the barke from the Tree, in any part thereof.

How to cutte your greate braunches, and when.

AND for the better remedie: First you shall cutte the same greate braunches, halfe a foote from the Tree, and after to sawe the reste cleane hard by the bodie of the Trée, then with a broad Chisell, cutte all cleane and smoth vppon that place, then couer it with Oxe dunge. Ye maie also cutte theim well in Winter, so that ye leaue the Trunke or braunche somewhat longer, so as ye maie dresse and cut theim againe in Marche and Aprill, as is before mentioned.

How ye ought to leaue these greate braunches cutte.

Ther thynges here are to bee shewed, of certaine greate and olde Trees onely; whiche in cuttyng the greate braunches thereof truncheon wise, doe renewe againe, as Walnuttes, Mulberie Trees, Plum Trées, Cherie Trees with others, whiche ye muste dissbraunche the bowes thereof, euen after Alhallowtide, or as sone as their leaues bee falne of; and likewise before thei beginne to enter

J.ij.   into

into Sappe.

Of Trees hauyng greate braunches.

THE saied greate braunches, when ye shall disbraunche theim, ye shall so cutte theim of in suche Truncheons, of lengthen the Tree, that the one maie bee longer then the other, that when the Cions bée growne good and long thereon, ye maie Graffe on theim againe as ye shall sée cause, accordyng as euery arme shall require.

Of barrennesse of Trees, the tyme of cuttyng ill braunches, and of vncoueryng the rootes.

SOmetymes a manne hath certaine olde Trees, whiche bée almoste spent, as of the Peare Trees, and Plum Trees, and other greate Trees, the whiche beare scant of fruite: but when as ye shall sée some braunches well charged therewith, then ye ought to cutte of all the other ill braunches and bowes, to the ende that those that remaine, maie haue the more Sappe to nourishe their fruite, and also to vncouer their rootes after Alhallowtide, and to cleaue the moste greatest rootes thereof (a foote from the tronke) and put into the saied cleftes, a thinne slate of hard stone, there lette it remaine, to the ende that the humour of the Tree, maie enter out thereby, and at the ende of Winter, ye shall couer hym againe, with as good and fatte yearth as ye can get, and let the stone alone.

¶ Trees the whiche ye must helpe, or plucke vp by the rootes.

ALL sortes of Trees whiche spryng Cions from the rootes, as Plum Trees, all kinde of Cherie Trees, and small Nutte Trees, ye muste helpe in pluckyng their Cions from the rootes in Winter, as sone as conueniently ye can, after the leafe is fallen. For thei dooe greatly plucke doune and weaken the saied trees, in drawyng to them the substaunce of the yearth.

What doeth make a good Nutte.

But

BVT chiefly to plant thefe Cions, the befte waie is to let them growe, and bée nourifhed twoo oʒ thʒée yeres from the roote, and then to tranfplant them, oʒ fette them in the Winter, as is afoʒefaied. The Cions whiche bee taken from the foote of the Hafell Trees, make good Nuttes, and to bée of muche ftrength and vertue, when thei are not fuffered to growe too long from the roote, oʒ foote afoʒefaied.

¶ Trees eaten with beaftes, muft
bee Graffed againe.

WHen certaine Graffes beyng well in Sappe, of thʒée oʒ fower yeres, oʒ there aboutes bee bʒo-ken, oʒ greatly endamaged with beaftes, whiche haue bʒoken thereof, it fhall little pʒofite to leaue thofe Graf-fes fo, but it were better to cutte them, and to graffe them higher, oʒ lower then thei were befoʒe. Foʒ the Graffes fhall take as well vppon the newe, as old Cion beyng graffed, as on the wilde ftocke: But it fhall not fo fone clofe, as vppon the wilde ftocke hedde.

How your wilde flockes ought not
haftely to bee remoued.

IN the beginnyng, when ye haue graffed your graf-fes on the wilde ftocke, doe not then haftely plucke vp thofe Cions, oʒ wilde ftockes fo Graffed, vntill ye fhall fée the Graffes put foʒthe a newe fheute, the whiche remainyng ftill, ye maie Graffe thereon againe, fo that your Graffes in haftie remouyng, maie chaunce to dye.

¶ When ye cutte of the naughtie
Cions from the Wood.

WHen your Graffes on the ftockes, fhall put foʒthe of newe Wood, oʒ a newe fheute, as of twoo oʒ thʒée foote long, and if thei put foʒthe alfo of other fmal fuperfluous Cions (about the faied members, oʒ bʒaunches that ye would nourifhe) cutte of all fuche ill Cions, harde by the hedde, in the fame yere thei are Graffed in, but not fo long as the Wood is in Sappe, till the Winter after.

¶ *How sometymes to cutte the*
principall members.

ALso it is good to cutte some of the pzincipall members oz bzaunches, in the firste yere, if thei haue too
many, and then againe, within twoo oz thzee yeres
after, when thei shall bee well spzong vp, and the Graffes wel
closed on the hedde of the storke: ye maie trimme and dzesse
theim againe, in takyng awaic the superfluous bzaunches, if
any there remain, foz it is sufficient inough to nourishe a yong
Tree, to leaue hym one pzincipall member on the hedde, so
that he maie bee one of those, that hath been Graffed on the
Tree befoze, yea, and the Tree shalbe fairer, and better in the
ende, then if he had twoo oz thzee bzaunches, oz pzecidence at
the foote. But if the tree haue been Graffed with many great
Cions, then ye must leaue hym moze largely, accozdyng as
ye shall see cause oz neede, to recouer the clestes on the hedde
of the saied graffe oz stocke.

¶ *How to guide and gouerne*
the saied Trees.

VVhen that your Trees doe beginne to spzing, ye must
ozder and see to them well, the space of thzee oz foure
yeres, oz moze, vntill thei bee well and strongly growne, in
helpyng theim aboue, in cuttyng the small twigges, and superfluous wood, vntill thei bee so high without bzaunches, as
a manne, oz moze if it maie bee, and then see to theim well, in
placyng the pzincipall bzaunches, if neede bee, with fozkes oz
wandes pzicke right, and well about them at the foote, and to
proine them, so that one bzaunche do not appzoche to nigh the
other, noz yet frete the one the other, when as thei doe enlarge
and growe, and ye must also cut of certaine bzaunches in the
Tree, where as thei are too thicke.

A kinde of sicknesse in Trees.

Hen certaine Trees are sicke of the Gall, whiche is
a kinde of sicknesse that doeth eate the barke, therefoze ye must cut it, and take out al the same infection
with a Chesill, oz suche like thyng. This must be doen at the
ends

ende of Winter, then put on that infected place of Dre dung,
oz Hogges dung, and binde it faſt thereon with cloutes, and
wzappe it with Dziars, ſo let it remaine a long tyme, till it
ſhall recouer againe.

Trees whiche haue wormes in the barke.

Of Trees whiche haue wozmes within their barkes,
is whe. e as pe ſhall ſee a ſwellyng oz riſing therein,
therfoze pe muſt cut oz cleaue the ſaied barke vnto the
wood, to the ende the humoz maie alſo diſtill out thereat, and
with a little hooke pe muſt pluckc oz dzawe out the ſaied wozmes, withall the rotten wood that pe can ſee, then ſhall pe put
vpon the ſaied place, a plaiſter made of Dre dung, oz of Hogs
dung medled and beaten with Sage, and a little of vnſleckt
Lime, then let it be all well blend together, and wzappe it on
a cloth, and binde it faſt and cloſe theron ſo long as it will hold.
The Lees of Wine ſhed oz poured vpon the rootes of Trees
(the whiche bee ſomewhat ſircke thzough the coldneſſe of the
pearth) whiche Lees doeth them muche good.

¶ Snayles, Antes and Wormes, doeth marre Trees.

Alſo pe muſt take heede of all maner of pong Trees: and
ſpecially of thoſe graffes, the whiche many wozmes and
Flyes doe endomage and hurt in the tyme of Sommer, thoſe
are the Snailes, the Piſmiars, oz Antes: the fielde Snaile
whiche hurteth alſo all other ſoztes of Trees that be greate,
pzincipally in the tyme that the Cuckowe doeth ſing, and betwixt Apzill and Midſomer, while thei bee tender. There bee
little beaſtes called Sowes, whiche haue many legges: and
ſome bee of them graie, ſome blacke, and ſome hath a long
ſharpe ſnowte, whiche bee very noyſome, and greate hurters
of pong Graſſes, and other pong Trees alſo, foz thei cut of in
eating the tender toppes (of the pong Cions) as long as ones
finger.

How ye ought to take the
ſaied Wormes.

F OR to take them well, ye muſt take heede and watche in the heate of the daie (your young Trees) and where ye ſhall ſee any, put your hand ſoftly vnderneath, without ſhaking the Tree, for thei wil ſodainly fall when one thinkes to take them : therefore ſo ſone as you can (that thei ſlipe not awaie no? fall) take him (quickly on the Cion) with your other hande.

To keepe Antes from yong Trees.

FOR to keepe the young Trees from Snailes and Antes : it ſhall be good to take Aſhes and to mingle vnſleckt Lime, beaten in pouder therewith, then lay it all about the roote of the tree, and when it raineth, thei ſhall be beaten downe into the Aſhes and dye : but ye muſt renewe your Aſhes after euery Raine from tyme to tyme : alſo to keepe them moyſt, ye muſt put certaine ſmall veſſelles full of water, at the foote of your ſaied Trees, and alſo the Lees of Wine, to be ſpread on the grounde there all aboutes. For the beſt deſtroying of the ſmall Snailes on Trees, ye muſt take good heede in the Spring tyme before the Trees bee leaued, then if ye ſhall ſee as it were ſmall wartes, knobbes o? braunches on the Trees, the ſame will bee Snailes. Prouide to take them awaie faire and ſoftly, before thei be full cloſed, and take heede that ye hurte not the wood o? barke of the ſaied Tree, as little as ye can, then burne thoſe braunches on the pearth, o? all to tread them vnder your feete, and then if any doe remaine o? renewe, looke in the heate of the daie, and if ye can ſee any, whiche will commonly be on the cleſtes o? for-kes of the braunches, and alſo vpon the braunches lying like tuſſes o? troopes together, then wrappe your handes all ouer with olde clothes, (and binde of leaues beneath them, and aboue them) and with your two hands rubb them downe there-in, and ſtraight waie ſire it, if ye doe not quickly with diligence thei will fall, and if thei fall on the pearth, ye can not lightly kill them, but thei will renewe againe, theſe kinde of wormes are noyſome Flyes whiche bee very ſtraunge, therefore take heede that thei doe not caſt a certaine redneſſe on

your

your face and bodie, for where as there be many of them, thei bee daungerous:it is straunge to tell of these kinde of Wormes, if ye come vnder or emong the Trees whereas bee many, thei will cast your face and handes, (your couered bodie, as your necke, breast and armes) full of small spottes, some redde, some blacke, some blewishe, whiche will so tingle and trouble you like Nettles, sometymies for a daie, or a daie and a night after:thei bee most on Plum trees, and Apple trees, nigh vnto moyst places, and ill apres: yet neuerthelesse, by the grace of God there is no daunger (that I vnderstande)to be taken by them. Ye shall vnderstande, that if it bee in the euenyng, or in the morning, when it raineth, thei will remaine about the graffyng place of the Tree; therefore it will be harde to finde them, because thei are so small. Moreouer, if suche braunches doe remaine in the vpper parte of the bowes or tree, ye shall binde of drie strawe about the bowes all vnder, then with a wispe on a Poles ende, set fire on all, and burne them.

¶ A note in Spryng tyme of
Fumigations.

HEre is to be vnderstoode & noted, that in the Spring tyme onely when Trees doe beginne to put foorth leaues and Blossomes, ye must then alwaies take heede vnto them, for to defende them from the Frost (if there come any) with Fumigations or smokes, made on the winde side of your Orchardes, or vnder your Trees)with Strawe, Hey, drie Chaffe, drie Ore dung, of Sawdust dried in an Ouen, of Tanners Ore dried likewise, of *Galbanum*, of olde shooes, Thatche of houses, of haire and suche like, one of these to be blende with an other:all these be good against the Frost in the Spryng tyme, and specially good against the East winde, which breedeth(as some saie)the Caterpiller worme.

To defende the Caterpiller.

AND some doe defende their Trees from the Caterpiller when the blossoming tyme is drie(if there be no Frost)by castyng of water, or salte water, euery se-

K.j.   conde

conde oɀ thirde daie vppon their Trees, (with Inſtrumentes foɀ the ſame, as with Squirtes of Wood oɀ Bɀaſſe oɀ ſuche like) foɀ in keepyng of them moyſt , the Caterpiller can not bɀeede thereon: this experience haue I knowne pɀoued of late to be good . Foɀ to conclude, he that will ſett oɀ plant Trees, muſt not paſſe foɀ any paines, but haue a pleaſure and delight therein, in rememeɀing the great pɀofite that commeth thereby: Againſt ſcarceneſſe of Coɀne , fruite is good ſtaie foɀ the pooɀe , and often it hath beene ſeene, one Aker of Oɀcharde grounde wooɀth fower Aker of Wheate grounde.

FINIS.

¶Here

¶ Here followeth a little treatise, how one
maie Graffe and Plant, subtile or Arti-
ficially, and to make many thynges
in Gardens verie straunge.

FOR to Graffe a subtill waie, take one oplet or
eye of a Graffe, slitt it rounde, aboue and beneath, and then behinde doune right, then
wreathe hym of, and sette hym vpon an other
Cion, as greate as he is, then dresse hym, as
is aforesaied, and he shall growe and beare.

To Graffe one Vine vpon an other.

FOR to Graffe on Vine vppon an other, ye shall cleaue
hym as ye dooe other Trees, and then putte the Vine
graffe in the clefte, then stoppe hym close and well with
Waxe, and so binde hym, and he shall growe.

If a Tree be too long without fruite.

YE shall vncouer his roote, and make a hole with a
Piercer, or small Auger, in the greatest roote he hath,
without pearcyng through the roote, then put in a pin
(in the saied hole) of drie wood, (as Oke or Ashe) and so let it
remaine in the saied hole, and stoppe it close againe with waxe
and then caste yearth and couer hym againe, and he shall beare
the same yere.

¶ For to haue Peaches twoo mo,
nethes before other.

R.ij.   Take

Take your Cions of a Peache Tree that dooeth soone
blossome in the Spryng tyme, and graffe them vppon
a frauke Mulberie tree, and he shall bryng of Peaches
twoo Monethes before others.

To haue Damsons or other Plummes, vnto Alhallowtide.

For to haue Damsons all the Sommer long, vnto Alhallowtide, and of many other kinde of sortes likewise,
ye shall graffe them vppon the Gooseberie tree, vppon
the frauke Mulberie tree, and vppon the Cherie tree, and thei
shall endure on the trees till Alhallowtide.

To make Medlars, Cheries and Peaches in eatyng, to tasse like spice.

For to make Medlars, Cheries, and Peaches, to taste
in the eatyng pleasaunt like spice, the whiche maie also
keepe vnto the newe come againe: ye shall graffe them
vppon the frauke Mulberie tree, as I haue afore declared, and
in the graffyng, ye shall wette them in Honie, and put a little
of the pouder of some good Spices, as the pouder of Cloues,
of Cinamon, or Ginger.

To make a Muscadell taste.

For to make a Muscadell taste, take a Gouge or Chesill of Iron, (and cutte your Sappe rounde about) then
putte in your Gouge or Chesill, vnder your Sappe on
your Cion, and raise three eyes or oylettes rounde about, and
so take of faire and softly your barke rounde aboute, and when
he is so taken of, dooe annointe it all ouer within the barke,
with pouder of Cloues, or Nutmegges, then set it on againe,
and stoppe it close with Waxe rounde aboute, that no water
maie enter in, and within thrice bearyng, thei shall bryng a
faire Muscadell Reison, whiche ye maie after bothe Graffe
and Plant, and thei shall bee all after a Muscadell fruite: some
slittes the barke doune, and so put in of spice.

To sette Apples and Peares, to come without blossomyng.

FOR to make Apples, and Peares, and other soꝛtes of fruite to come without blossompng, that is, ye shall Graffe them (as ye dooe other kinde of fruite) vppon the Figge Tree.

To haue Apples and Chestnuites rathe, and also long on the Trees.

FOR to haue Apples called(in Frenche)*de blanc Durell,* oꝛ *de Yroael,* and of Chestnuttes very rathe, and long(as vnto Alhallowtide)on the trees:and to make suche fruite also to endure, the space of twoo yeres, ye shall graffe them on a laterward fruite, as Pome Richard, oꝛ vpon a Peare Tree, oꝛ Apple tree of *Dangoiſſe.*

¶*To haue good Cheries on the Trees, at Alhallowtide.*

TO haue Cheries on many trees, good foꝛ to eate vn= to Alhallowetide,ye shall Graffe them vpon a franke Mulberie tree, and likewise to Graffe them vppon a Willowe, oꝛ Sallowe tree, and thei shall endure vnto Alhal= lowtide on the trees.

¶*To haue rathe Medlars twoo monethes before others.*

FOR to haue Medlars twoo Monethes souer then others: and that the one shalbe better farre then the other, ye shall Graffe them vppon a Gooseberie tree, and also a franke Mul= berie tree, and befoꝛe ye doe Graffe them, ye shall wet them in Honie, and then Graffe them.

For to haue rathe or tymely

Peares.

FOR to haue a rathe Peare, the whiche is in Fraunce, as the Peare *Cailonet,* and the Peare *Haſtinean.* Foꝛ to haue them rathe oꝛ sone,ye shall Graffe them on the Pine tree: And foꝛ to haue them late,ye shall Graffe them on the Peare, called in Fraunce *Dangoiſſe,* oꝛ on other like hard Peares.

To haue Miſples or Medlars without ſtones.

K.iij.   Foꝛ

OR to haue Medlars without ſtones, the whiche ſhal
F taſte ſweete as Honie, ye ſhall Graffe them as the o-
ther, vpon an Eglentine, o2 ſweete B2iár tree, and ye
ſhall wette the Graſſes (befo2e ye Graffe them) in Honie.

To haue Peares betymes.

OR to haue the Peare of *Angwiſſe*, o2 *Permain*, o2
F *Satigle*, (whiche bee of certaine places ſo called) a Moneth o2 twoo befo2e others, the which ſhall endure, and
be good vnto the newe come againe, ye ſhall graffe them vpon
a Quince tree, and likewiſe vpon the franke Mulberie tree.

¶ *To haue ripe or franke Mulberies*
verie ſone and late.

OR to haue franke o2 ripe Mulberies very ſoone, ye
F ſhall graffe them vpon a rathe Peare tree, and vpō the
Gooſeberie tree, and to haue very late, and to endure
vnto Alhallowtide, ye ſhall graffe them vpon the Medlar tree.

To keepe Peares a yere.

NOw fo2 to keepe Peares a yere: ye ſhall take of fine
Salt verie d2ie, and put thereof with your Peares into a Barrell, in ſuche ſo2t, that one Peare do not touch
an other, ſo fill the Barrell if ye liſte, then ſtoppe it, and let it
bee ſet in ſome d2ie place, that the Salt doe not waxe moiſte,
thus ye maie keepe them long and good.

¶ *To haue your fruite taſte halfe*
Apples, halfe Peares.

IF ye will haue your fruite táſte halfe a Peare, and half an
Apple, ye ſhall in the ſp2ing take graffes, the one a Peare,
and the other an Apple, ye ſhall cleaue o2 pare them in the
graffyng ioynt o2 place, and ioyne halfe the Peare Cion, and
ſo ſet them into your ſtocke, and ſee well that no raine doe enter therein vppon your ioynt, and that fruite ſhall b2yng thee
halfe a Peare, and the other halfe an Apple in taſte.

Tymes of Graffyng.

IT is good alſo to Graffe one o2 twoo daies befo2e the
chaunge, and no mo2e, fo2 looke ſo many mo2e daies, as ye
ſhall Graffe befo2e them, ſo many mo yéres it will be, ere
your

your trees ſhall bꝛing fruite:alſo it is good graffing all the increaſe of the Moone,but the ſoner after the chaſnge,the better.

To graffe the Quine Apple.

IF ye graffe the Quine Apple, vppon an Apple ſtockꝛ,he
ſhall not long continue without the Canker , but to graffe
hym on a knottie young Crabſtocke , he ſhall indure long
without the Canker.

❡ To deſtroye Piſmiars or Antes, about a Tree.

FOꝛ to deſtrope Emicts oꝛ Antes,whiche be about a tree,
if ye remoue and ſtirre the pearth all about the roote of the
ſaied tree,then put thereon all about,a greate quantitie of the
Soote of a Chimney,and the Antes oꝛ Piſmiars will either
awaie,oꝛ els ſhoꝛtlp dye.

An other for the ſame.

ALſo an other waie foꝛ to deſtrop Antes is,pe ſhall take
of the Sawduſt of Oke wood onelp , and ſtrowe that
all about the Tree roote, and the next raine that doeth
come , all the Piſmiars oꝛ Antes ſhall dpe there : Foꝛ Eare-
wigges,ſhooes ſtopt with Haie,and hanged on the Tree one
night,thei come all in.

❡ To haue Nuttes, Plummes, and Almondes.

❡ Nuttes greater then others.

FOꝛ to haue greate Nuttes,Plums,and Almondes greater thē others,ye ſhall take fower Nuts,
oꝛ of any of this fruite aboue ſaied, and put them
into a pott of pearth , iopnyng the one to the other
as nere as ye can, then make a hole in the bottome of the pot,
thꝛough the whiche holes, theſe Nuttes ſhall bee conſtrained
to iſſue , and beyng ſo conſtrained , ſhall come to perfection
and growe togethers as in one Tree,the which in tpme ſhall
bꝛyng his fruite moꝛe greater and larger,then others.

❡ To make an Oke or other Tree greene in Winter as in Sommer.

Alſo

ALſo to make an Oke oꝛ other Tree to bee greene as well in Winter as in Sommer, ye ſhall take the Graffe of an Oke Tree, oꝛ other Tree, and graffe it vpon the Hollie tree: the beſt and moſt ſureſt waie is, to graffe one thꝛough the other. Alſo who ſo will edifie oꝛ make an Oꝛcharde, he ought (if he can) to make it in a moyſt place, where as the South windes, oꝛ Sea windes maie haue recourſe vnto them.

¶ The tyme of plantyng without rootes,
and with rootes.

ALſo the beſt tyme to plant oꝛ ſett without rootes as with bꝛaunches oꝛ ſteuerynges of all ſoꝛtes of Trees whiche hath a greate pithe, as Figge trees, Haſell trees, Mulberie trees, and Vines, with other like Trees, all whiche ought to be ſet from the middeſt of September (if the leaues be of) vnto Alhallowtide, and all other Trees with rootes, ought to be ſet in Aduent vntill Chꝛiſtmas, oꝛ anone after, if the tyme be not very colde and daungerous.

To keepe fruite from the Froſt.

ALſo to keepe fruite from the Froſt, and in good colour, vnto the new come again, ye ought ſo foꝛ to gather them when the tyme is faire and dꝛie, and the Moone in her decreſing, and that thei lye alſo in very dꝛie places by night, couered thinne with Wheate ſtrawe, and if the tyme of Winter be colde and very harde, then put of Haie aboue them in your Strawe, and take it awaie when as a faire tyme commeth: and thus ye ſhall keepe your fruite faire and good.

The daies to plant and graffe.

ALſo (as ſome ſaie) from the firſt daie of the new Moone, vnto the thirteene daie thereof, is good foꝛ to Plant, oꝛ Graffe, oꝛ ſowe, and foꝛ greate neede, ſome doe take vnto the ſeuenteene oꝛ eighteene daie thereof, and not after, neither graffe noꝛ ſowe, but as is afoꝛe mencioned, a daie oꝛ two daies afoꝛe the chaunge, the beſt ſignes are, Taurus, Virgo, and Capꝛicoꝛne.

To haue greene Roſes all the yere.

Foꝛ

Days of y͏ᵉ moon
is to graft

FO2 to haue greene Roses, ye shall (as some saie) take your
Rose buddes in the Spryng tyme, and then graffe them
vpon the Hollie stocke, and thei shall bee greene all the yere.

To keepe Reisons or Grapes good a yere.

FO2 to keepe Reisons or Grapes good all a whole yere,
ye shall take of fine drie Sande, and then laie pour Reisons or Grapes therein, and it shall keepe them good a
whole yere. Some keepe them in a close glasse from the ayre.

To make fruite laxatiue from the Tree.

FOR to make any fruite laxatiue from the Tree, what
fruite soeuer it be, make a hole in the stock, or in the maister roote of the tree, (with a greate Pearcer slope wise)
not through, but vnto the pithe, or somewhat further, then fill
the saied hole with the iuyce of Elder, of *Centorie*, of *Seney*, or
of *Turbith*, or suche like laxatiues, then fill the saied hole therwith of whiche of them ye will, or els ye maie take three of
them togethers, and fill the saied hole therewith, and then stop
the saied hole close with soft Waxe, then claie it thereon, and
put Mosse very well ouer all, so that nothyng maie issue or
fall out, and all the fruite of the saied Tree shalbe from thence
forth laxatiue.

A note for all Graffers and Planters.

ALso whensoeuer ye shall Plant or Graffe, it shall bee
meete and good for you to saie as followeth. In the
name of GOD the Father, the Sonne and the holie
Ghost, Amen. Increase and multiplie, & replenishe the earth:
and saie the Lordes prayer, then saie: Lord God heare my pra-
per, and let this my desire of thee be hearde. The holie spirite
of God which hath created all thynges for man, and hath ge-
uen them for our comfort. in thy name O Lord we set, plant,
and Graffe, desiryng that by thy mightie power thei
maie encrease, & multiplie vpon the earth, in bea-
ring plentie of fruite, to the profite and comfort of all thy faithfull people, through
Christ our Lorde, Amen.
...FINIS.

L.j.   ¶ Here

¶ Here followeth certaine waies of Plan-
tyng and Graffyng , with other necessaries
herein meete to be knowne, translated
out of Dutch by L. M.

To graffe one Vine on an other.

YOU that will graffe one Vine vpon an other,
ye shall (in Ianuarie) cleaue the head of the
Vine, as ye doe other stockes, and then put in
your Vine Graffe or Cion , but first ye must
pare him thinne, ere ye set him in the head, then
Claie and Mosse him as the other.

Chosen daies to graffe in, and to choose your Cions.

ALso whensoeuer that ye will Graffe , the best chosen
tymes is on the last daie before the chaunge, and also
in the chaunge, & on the second daie after the change,
if ye graffe (as some saie) on the thirde, fourth and fift daie after the chaunge , it will bee so many yeares ere those Trees
bzyng

bzyng fozthe fruite. Whiche thyng ye maie beleeue if ye will, but I will not. Foz some doe holo opinion, that it is good graf-fyng from the chaunge, vnto the rviij. daie thereof, whiche I thinke to bee good in all the increasyng of the Moone, but the soner the better.

¶ To gather your Cions.

ALso suche Cions oz Graffes, whiche ye doe gett on the other Trees, the yong Trees of thzee oz fower peres, oz fiue oz sire peres are beste to haue Graffes. Take them of no vnder bowes, but in the topp vpon the East side, if ye can, and of the fairest and greatest. Ye shall cut them twoo inches long of the olde Wood, beneath the ioynte. And whensoeuer ye will Graffe, cutte oz pare your graffes taper-wise from the ioynt, twoo inches oz moze of length, whiche ye shall set into the stocke: and befoze ye set it in, ye shall opē your stocke with a wedge of Iron, oz harde wood, faire and softly: then if the sides of your clestes bee ragged, ye shall pare theim with the pointe of a sharpe knife on bothe sides, within and a-boue, then set in your graffes close on the outsides, and also a-boue: but let your stocke be as little while open as ye can, and when your graffes bee well set in, plucke fozthe your wedge: and if your stocke do pinche your graffes muche, then ye must put in a wedge of the same wood to helpe your graffes: Then ye shall laye a thicke barke oz pill ouer the cleft, from the one grafte to the other, to keepe out the claie and raine, and so claie them twoo fingers thicke rounde aboute the cliffes, and then laie on Mosse, but Wooll is better nert to your claie, oz els to temper your claie with Wooll oz Haire, foz it shall make it bide closer. and also stronger on the stocke hedde. Some take Wooll nerte the claie, and wzappeth it all ouer with Linnen cloutes: foz the Wooll beyng once moiste, will keepe the claie so a long tyme. And other some take Woollen cloutes, that haue been laied in the iuice of Wozmewood, oz suche like bit-terthyng, to keepe creepyng Wozmes from commyng vn-der to the Graffes. If ye graffe in Winter, put your claie vp-permoste, foz Somer your Mosse. Foz in Winter the Mosse

is warme,and pour claie will not cleaue. In Sommer your
claie is colde, and your Mosse keepes hym from cleauyng or
chappyng. To binde them,take of Willowe pilles,of clouen
Briers,of Oziers,or suche like. To gather your Graffes on
the Easte parte of the tree is counted beste:if ye gather them
belowe on the vnder boughes, thei will growe flaggie, and
spreadyng abroade: If ye take them in the toppe of the tree,
thei will growe vpright.Yet some doe gather their Cions or
Graffes on the sides of the trees,and so graffe them againe on
the like sides of the stockes, the whiche is of some menne not
counted so good for fruite. It is not good to graffe a greate
stocke,for thei will be long ere thei couer the hedde thereof.

Of Wormes in Trees or fruite.

IF ye haue any trees eaten with Wormes, or doe bryng
Wormie fruite, ye shall vse to washe all his bodie and
greate braunches,with twoo partes of Cowpisse,and one
part of Uineger, or els if ye can get no Uineger, with Cowe-
pisse alone, tempered with common Ashes: then washe your
trees therewith before the Spryng,and in the Spryng,or in
Sommer. Anniseedes sowne about the tree rootes, driue a-
waie Wormes,and the fruite shalbe the sweeter.

The settyng of Stones,and ordryng thereof.

AS for Almonde trees, Peache trees, Cheric trees,
Plum trees,or others,ye shall thus plant or set them.
Laie first the Stones in water,three daies and foure
nightes,vntil thei sinke therein:then take them betwixt your
finger and your thumbe, with the small ende vpward,and so
set them twoo fingers deepe in good pearth.And when ye haue
so doen, ye shall rake them allouer, and so couer them: and
when thei beginne to growe or spring,keepe them from weedes,and thei shall prospere the better,specially in the first yere.
And within twoo or three yeres after, ye maie set or remoue
them where ye liste, then if ye doe remoue them againe after
that,ye muste prolue of all his twigges,as ye shall see cause,
nigh the stocke:thus ye maie doe of all kinde of trees,but specially those whiche haue the greate Sappe, as the Mulberie,

oʒ Figge tree, oʒ suche like.

To gather Gumme of any Tree.

IF ye liſte to haue the Gumme of an Almond tree, ye ſhall ſticke a greate naile into the tree, a good waie, and ſo lette hym reſt, and the Gumme (of the tree) ſhall iſſue thereat: thus doe menne gather Gumme of all ſoʒtes of trees: yea, the common Gumme that men doe vſe and occupie.

To ſet a whole Apple.

ALſo ſome ſaie, that if ye ſet a whole Apple fower fingers in the pearth, all the Pepines oʒ Curnelles in the ſame Apple, will grow vp togethers in one whole ſtocke oʒ Cion, and all thoſe Apples ſhalbee muche fairer and greater then others: but ye muſt take heede, how ye doe ſette thoſe Apples, whiche doe come in a Leape pere, foʒ in a leape pere (as ſome doe ſaie) the Curnelles oʒ Pepines, are turned contrary, foʒ if ye ſhould ſo ſett, as commonly a manne doeth, ye ſhall ſet them contrary.

Of ſettyng the Almonde.

ALmondes doe come foʒthe and growe commonly wel if thei be ſet without the ſhell oʒ huſke, in good yearth oʒ in rotten Hogges dunge: If ye laie Almondes one daie in Uineger, then ſhall thei (as ſome ſaie) be very good to Plant, oʒ laie hym in Milke and water, vntill he doe ſinke, it ſhalbe the better to ſet, oʒ any other Nutte.

Of Pepines watered.

THe Pepines and Curnels of thoſe trees, whiche haue a thicke oʒ rough barke, if ye laie them thʒee daies in water, oʒ els vntill thei ſinke therein, thei ſhall bee the better, then ſet them, oʒ ſowe them, as is afoʒe mencioned, and then remoue them, when theibe well rooted, of thʒee oʒ foure yeres grouth, and thei ſhall haue a thinne barke.

To Plant or ſet Vines.

IF ye Plant oʒ ſet Uines, in the firſt oʒ ſecond yere, thei wil bʒyng no fruite, but in the third yere thei will beare, if thei be well kept: ye ſhall cut them in Januarie, and ſet them ſone after thei be cut from the Uine, and ye ſhall ſet twoo together

L.iij.   the

the one with the old wood, and the other without, and so lette
them grow, in pluckyng awaie all weedes from about them,
and when ye shall remoue them in the second and third yere,
beyng well rooted, ye shall set them well a foote depe (in good
fat pearth) with good doung, as of one foote depe, or there a-
boutes, and kepe them cleane from weedes, for then thei will
prosper the better, and in Sommer when the Grape is knitt,
then ye shall breake of his toppe or braunche, at one or twoo
iopntes after the Grape, and so the Grape shalbe the greater,
and in the Winter when ye cut theim, ye shall not leaue paste
twoo or three leaders on eche braunche, on some braunche but
one leader, whiche must bee cut betwixt twoo iopntes, and ye
shall leaue the pong Cline to be the leader: Also ye shall leaue
thereof three or fower iopntes at all tymes, if a pong Cion do
come forthe of the old brauche, or side thereof, if ye doe cut him,
ye shall cut hym hard by the old braunche, and if ye will haue
hym to bryng the Grape nexte yere, ye shall leaue twoo or
three iopnts thereof, for the pong Cion alwaies bryngeth the
Grape: ye maie at all tymes, so that the Grape be once taken
and knit, euer as the superfluous Cions doe growe, ye maie
breake theim of at a iopnt, or hardly by the old braunche, and
the grapes will be the greater: this ye maie order your Cline
all the Sommer long without any hurte.

To set or plant the Cherie.

Cherie Trees, and all the Trees of stone fruite,
would be planted or set of Cions, in colde groundes
and places of good pearth, and likewise in high or
hillie places, dry and well in the shade: if ye doe remoue, ye ought to remoue them in Nouember and Januarp.
If ye shall see your Cherie Tree waxe rotten, then shall ye
make a hole in the middest of the bodie twoo foote aboue the
grounde, with a bigge Wearcer, that the humour maie passe
forthe thereby, then afore the Spryng, shutt hym vp againe
with a pinne of the same Tree: thus ye maie doe vnto all o-
ther sortes of Trees when thei beginne to rotte, and is also
good for them whiche beare scant of fruite or none.

T.

To keepe Cheries good a yere.

OR to keepe Cheries good a yeare, ye shall cutt of the stalkes, and then laie them in a well leaded potte, and fill the said pot therewith, then put vnto them of good thin Honey, and fill the saied potte therewith, then stoppe it with Claie that no aire enter in, then sett them in some faire Seller, and put of Sande vnder, and all aboue it, and couer the pot well withall, so let it stande or remaine: thus ye maie keepe them a pere, as freshe as though thei came from the Tree, and after this sorte ye maie keepe Peares, or other fruite.

Against Pismiars.

IF ye haue Cherie Trees laded or troubled with Pismiars or Antes, ye shall rubbe the bodie of the Tree, and all about the roote with the iuyce of Purslaine, mingled halfe with Uineger. Some doe vse to anoynt the Tree beneath all about the bodie, with Tarre and Bird lime, with Wooll oyle boyled together, and anoynt the Tree beneath therewith, and doe laie of Chalke stones all about the Tree roote, some saie it is good therefore.

The settyng of Chestnuttes.

HE Chestnut Tree, men doe vse to plant like vnto the Figge Tree. Thei maie bee bothe planted and graffed well, thei waxe well in freshe and fatt pearth, for in Sande thei like not: If ye will sett the Curnels, ye shall laie them in water vntill thei doe sincke, and those that doe sincke to the bottome of the water bee best to set, which ye shall set in the Moneth of Nouember and December, fower fingers deepe, a foote one from an other, for when thei be in these two Monethes set or planted, thei shall endure long, and beare also good fruite, yet some there be that plant or sett them first in dung, like Beanes, whiche will bee sweeter then the other sorte, but those whiche be set in the two Monethes aforesaied, shall first beare their fruite: Men maie proue whiche is best, experience doth teache. This is an other waie to proue and knowe, which Chestnuttes be best to plant or set, that is: ye shall take a quantie of Nuttes, then lay them

iij

in Sande the space of thirtie daies: then take and washe them in water faire and cleane, and throw them into water againe, and those whiche doe sincke to the bottome, are good to plant or set, and the other that swimme are naught: thus maie ye doe with all other Curnelles or Nuttes.

¶ To haue all stone fruite taste, as ye shall thinke good.

IF ye will haue all stone fruite taste as ye shall fansie or thinke good, ye shall first laie your stones to soke in suche licour or moysture, as ye will haue the fruite taste of, and then sette them: as for the Date tree (as some saie) he bringeth no fruite except he bee a hundred yeres olde, and the Date stone must soke one Moneth in the water before he bee sett, then shall ye set hym with the small ende vpwarde in good fatte yearth, in hott Sandie grounde sower fingers deepe, and when the bowes doe begin to spryng, then shall ye euery night sprinckle them with raine water, (or other if ye haue none) so long till thei become forth and growne.

Of graffyng the Medlar and Misple.

OR to graffe the Medlar or Misple: men doe vse to graffe them on the white Hathorne Tree, thei will proue well, but yet small and sower fruite: to graffe one Medlar vpon an other is the better, some men doe graffe first the Wildyng Cion vpon the Medlar stocke, and so when he is wel taken and growne, then thei graffe thereon the Medlar againe, the whiche doeth make them more sweete, very greate and faire.

Of the Figge Tree.

THE Figge Tree in some Countrey, beareth his fruite fower tymes a yeare, the blacke Figges are the best, beyng dried in the Sunne, and then laied in a vessell in beddes one by an other, and then sprinkled or strawed all ouer, euery laie with fine Meale, then stoppe it vp, and so it is sent out of the lande. If the Figge tree will not beare, ye shall digge him all about, and vnder the rootes in Februarie,

rie, and take out then all his earth, and put vnto hym the dung
of a P?iuie, fo? that he liketh best: ye maie mingle with it of o-
ther fatt pearth, as Pigeons dung mingled with Oyle and
Pepper stampt, whiche shall fo?warde him muche to nopnt
his rootes therewith: ye shall not plant the Figge tree in cold
tymes, he loueth hot, stonie, o? grauely grounde, and to bee
planted in Autumne is best.

Of the Mulberie Tree.

IF ye will plant the Mulberie Tree, the Figge Tree
o? others which b?ing no seede, ye shall cut a twigge
o? b?aunche (from the tree roote) of a peres grouth,
with the olde wood o? barke, about a cubite long, whiche ye
shall plant o? set all in the pearth, saue a shaftment long of it,
and so let it growe, in wateryng it as ye shall see neede. This
must be doen befo?e the leaues begin to sp?ing, but take heede
that ye cut not the ende o? toppe aboue, fo? then it shal wither
and d?ie.

Of Trees that beare bitter fruite.

OF all suche Trees as beare bitter fruite, to make them
b?yng sweeter, ye shall vncouer all the rootes in Januarie, and take out all that pearth, then put vnto them of Hogs
dung greate plentie, and then after putt vnto them of other
good pearth, and so couer them therewithall well againe, and
their fruite shall haue a sweeter taste. Thus men maie doe
with other Trees whiche b?yng bitter fruite.

To helpe barren Trees.

HEre is an other waie to helpe barren Trees, that thei
maie b?yng fruite: if ye see your tree not beare scantly
in th?ee o? fower peres good plentie, ye shall bo?e an
hole with an Auger o? Pearcer, in the greatest place of the
bodie, (within a parde of the grounde) but not th?ough, but
vnto o? past the harte, ye shall bo?e him a slope: then take honey and water mingled together a night befo?e, then put the
saied honey and water into the hole, and fill it therwith; then
stoppe it close with a sho?t pinne made of the same Tree, not
striken into farre fo? pearcyng the licour.

M.j.   An

An other waie.

IN the beginnyng of Winter, ye shall digge those Trees rounde about the rootes, and lett them so rest a daie and a night, and then put vnto them of good pearth, mingled well with good store of watered Otes, or with watered Barley or Wheate, laid next vnto the rootes, then fill it with other good pearth, and he shall beare fruite, euen as the boryng of a hole in the maister roote, and strike in a pinne, and so fill hym againe, shall helpe hym to beare, as afore declared.

To keepe your fruite.

ALL fruite maie be the better kept, if ye laie theim in drie places, in drie Strawe or Hape, but Hape ripeth to sore, or in a Barley mow, not touchyng one ye other, or in Chaffe, or in vessels of Iuniper, or Cipers wood, ye maie so kéepe them well in drie Salt or Honie, and vppon boardes, where as fire is nigh all the Winter, also hangyng nigh fire in the Winter, in Nettes of yarne.

The Mulberie Tree.

THE Mulberie tree, is planted or sette by the Figge tree, his fruite is first sower, and then swete, he liketh neither dewe nor raine, for thei hurt hym, he is well pleased with foule pearth and dung: His braunches will waxe drie within euery sixe yeres, then must ye cut them of, as for other trees, thei ought to be proined euery yere, as ye shall sée cause, and thei will be the better, and to plant hym from midde Februarie, to midde Marche is best.

Of Mosse of the Tree.

IF the Mosse on your Trees, ye must not lette it too long be vncleansed, ye must rubbe it of with a Grate of Wood, or a rough Haire, or suche like, in Winter when thei bee moiste or wet, for then it will of the souer, for Mosse dooeth take awaie the strength and substaunce of the fruit, and makes the trees barren at length: when you se your trees begin to waxe Mossie, ye must in the Winter vncouer their rootes, and put vnder them good pearth, this shall helpe them, and keepe them long without Mosse: for the pearth not
stirred

ſtirred aboue the roote, is one cauſe of moſſineſſe, and alſo the
barrenneſſe of the grounde whereon he ſtandeth, and your
Moſſe doeth ſucker in Winter, Flies and other Uermin, and
ſo doeth therein hide them in Sommer, whiche is occaſion of
eatyng the bloſſomes, and tender Cions thereof.

To keepe Nuttes long.

FOR to keepe Nuttes long, ye ſhall drie them, and
couer them in drie Sande, and put theim in a drie
Bladder, or in a Fatte made of Walnut Tree, and
put of drie Juie beries therein, and thei ſhall bee
muche ſweeter. To keepe Nuttes greene a yere, and alſo
freſhe, ye ſhall put them into a potte with Honie, and thei ſhall
continue freſhe a yere, and the ſaied Honie will bee gentle and
good for many Medicines. To keepe Walnuttes freſhe and
greene, in the tyme of ſtrainyng of Ueriuyce, ye ſhall take
of that Pommis, and put thereof in the bottome of a Barrell,
then laie your Walnuttes all ouer, then Pommis ouer them,
and ſo Walnuttes againe, and then of the Pommis, as ye
ſhall ſee cauſe to fill your veſſell. Then ſtoppe hym cloſe as ye
doe a Barrell, and ſet hym in your Seller, or other place, and
it ſhall keepe your Nuttes freſhe and greene a yere. Some
vſe to fill an yearthen potte with ſmall Nuttes, and then put
to theim drie Sande, and couer theim with a lidde of yearth,
or ſtone, and then thei claye it, ſettyng the mouthe of the potte
douneward, twoo foote within the yearth, in their Gardine,
or other place, and ſo thei will keepe verie moiſte and ſweete
vntill newe come.

To cut the Peache tree.

THE Peache tree is of this nature, if he bee cutte
(as ſome ſaie) greene, it will wither and drie.
Therefore if ye cutte any ſmall braunche, cutte
it harde by the bodie: the withered twigges euer
as thei wither, muſt bee cutte of hard by the greate braunche,
or bodie thereof, for then thei dooe proſpere the better. If a

Peache tree doe not like, ye shall put to his rootes, the Lees of Wine mingled with water, and also washe his rootes therewith, and likewise the braunches, then couer hym againe with good pearth mingled with his owne leaues, for those he liketh beste. Ye maie graffe Peache vpon Peache, vpon Hasill, or Ashe, or vppon Cherie Tree, or ye maie graffe the Almonde vpon the Peache tree. And to haue greate Peaches, ye must take Cowes milke, and putte good pearth thereto, then all to strike the bodie of the tree therewith, bothe vpward and rounward, or els open the roote all bare, three daies and three nightes, then take Goates milke, and washe all the rootes therewith, and then couer them againe: this muste bee dooen when thei beginne to blossome, and so shall he bryng greate Peaches.

To colour Peache stones.

TO colour Peache stones, that all the fruite thereof shall haue the like colour hereafter, that is: Ye shall lape or sette Peache stones in the pearth seuen daies or more, vntill ye shall se the stones beginne to open, then take the stones and the curnelles softly forthe thereof, and what colour ye will, colour the curnell therewith, and put them into the shell againe, then binde it faste together, and sette it in the pearth, with the small ende vpward, and so lette hym growe, and all the Peaches, whiche shall come of the same fruit (graffed or vngraffed) will be of the same colour. The Peache tree ought to be planted in Autumne, before the cold doe come, for he can not abide the cold.

❡ If Peache Trees bee troubled with Wormes.

ALso if any Peache tree bee troubled with Wormes, ye shall take twoo partes of Cowe pisse, with one part of Uineger, then shall ye sprinckle the tree all ouer therewith, and washe his rootes and braunches also, and it will kill the Wormes: this maie ye doe vnto all other trees, whiche be troubled with Wormes.

To

To haue the Peache without stones.

FOR to make the Peache growe without stones, ye shall take a Peache Tree newlie planted, then sette a Willowe hard by, whiche ye shall bore a hole through, then putte the Peache tree through the saied hole, and so close hym on bothe sides thereof, Sappe to Sappe, and let hym so growe one yere, then the next yere ye shall cutt of the Peache stocke, and lette the Willowe feede hym, and cutte of the upper part of the Willowe also three fingers high: and the nexte Winter sawe hym of nigh the Peache, so that the Willowe shall feede but the Peache onely: and this waie ye maie haue Peaches without stones.

An other waie for the same.

YE shall take the Graffes of Peaches, and Graffe them upon the Willowe stocke, and so shall your Peaches bee likewise without stones.

If Trees doe not prosper.

IF that ye see that your Trees doe not waxe nor prosper, take and open the rootes in the beginnyng of Januarie, or afore, and in the biggest roote thereof, make a hole with an Auger, to the pithe or more, then strike therein a pinne of Oke, and so stoppe it againe close, and lette it bee well waxte all aboute the pinne, then couer hym againe with good yearth, and he shall dooe well: some dooe use to cleaue the roote.

¶ *How to graffe Apples, to laste on the Tree till Alhallowtide.*

HOw ye maie haue many sortes of Apples uppon your Trees untill Alhallowtide, that is, ye shall Graffe your Apples uppon the Mulberie Tree, and upon the Cherie Tree.

¶ *To make Cheries and Peaches smell, and taste like spice.*

How to make that Cherries and Peares, shall bee pleasaunte, and shall smell and taste like spice, and that ye maie keepe them well, till the newe do come againe, ye shall Graffe them on the Mulberie Tree, as is aforesaied: But first ye shall soke them in Honie and Water, wherein ye shall putte of the pouder of Cloues, Ginger, and Cinamon.

⸿ To graffe an Apple whiche shall be halfe sweete, and halfe sower.

To graffe that your Apples shall bee the one halfe sweete, and the other halfe sower: ye shall take two Cions, the one sweete and the other sower, some doe put the one Cion through the other, and so graffes them betweene the barke and the Tree: and some againe doe pare bothe the Cions finely, and so settes them ioyning into the stocke, inclosing Sappe to Sappe, or bothe the outsides of the graffes, vnto the outsides of the stocke, and so settes them into the head as the other: and thei shall bryng fruite, the one halfe sweete, and the other halfe sower.

To graffe a Rose on the Holly.

For to graffe the Holly, that his leaues shall keepe all the yeare greene: Some doe take and cleaue the Holly, and so graffes in a white or redd Rose budde, and then puttes Claie and Mosse to hym, and lettes hym growe, and some doe put the Rose budde into a slitte of the barke, and so putteth Claie and Mosse, and bindes hym featelp therein, and lets him growe, and he shall carie his leafe all the yere.

Of keepyng of Plummes,

Of Plummes there bee many sortes, as Damsons, whiche bee all blacke, and counted the best: All maner of other Plummes a man maie keepe well a yeare, if thei be gathered ripe, and then dried, and put into vesselles of Glasse: If ye can not drie them well in the Sunne, ye shall drie them on Hurdelles of Oziars made like Lettice windowes, in a hot Ouen after Breade is drawne forthe, and so reserue

ſerue them. If a Plum tree like not, open his roote, and poo2e in all about the d2egges of Wine mixt with water, and ſo co∙ uer hym well againe, o2 poo2e on them ſtale Urine, o2 olde piſſe of olde men, mixt with twoo partes of water, and ſo co∙ uer hym as befo2e.

¶ Of alteryng of Peares, or ſtonie fruite.

IF a Peare doe taſte harde o2 grauely about the co2e, like ſmall ſtones, ye ſhall vncouer his rootes (in the Winter, o2 afo2e the Sp2yng) and take out all the pearth thereof, and picke out all the ſtones as cleane from the pearth as ye can about his roote, then ſift that pearth, o2 els take of other good fatte pearth without ſtones, and fill all his rootes againe therewith, and he ſhall b2yng a ſoft and gentle Peare to eate, but ye muſt ſee well to the wateryng of hym often.

The making of Cyder and Perrie.

IF Apples and Peares, men doe make Cyder and Perrie, and becauſe the vſe thereof in moſt places is knowne, I wil here let paſſe to ſpeake any fur∙ ther thereof, but this (in the p2eſſyng your Cy∙ der) I will counſell you to keepe cleane your veſſelles, and the places whereas your fruite doeth lye, and ſpecially after it is b2uſed o2 b2oken, fo2 then thei d2aw filthie aire vnto them, and if it bee nigh, the Cyder ſhall bee infected therewith, and alſo beare the taſte after the infection thereof: therefo2e as ſoone as you can, tunne it into cleane and ſweete veſſelles, as into veſſelles of white Wine, o2 of Sacke, o2 Claret, and ſuche like, fo2 theſe ſhall keepe your Cyder the better and the ſtronger a long tyme after: Ye maie hang a ſmall bagge of linnen by a th2edd doune into the lower parte of your veſſell, with pouder of Cloues, Mace, Cinamon, and Ginger, and ſuche like, whiche will make your Cyder to haue a pleaſaunt taſte.

To helpe frofen Apples.

Of

Of Apples that bee frosen in the colde and extreme Winter. The remedie to haue the Ise out of them, is this. Ye shall laie them first in colde water a while, and then laie them before the fire, or other heate, and thei shall come to themselues againe.

To make Apples fall from the Tree.

If ye put of fierp coles vnder an Apple Tree, and then cast of the pouder of Brimstone therein, and the fume thereof ascende vp, and touche anp Apple that is wet, that Apple shall fall incontinent.

To water Trees in Sommer, if thei waxe drie about the roote.

Whereas Apple Trees be set in drie grounde, and not deepe in the grounde, in Sommer if thei want moysture, ye shall take of Wheate straw, or other, and euerp euenyng (or as ye shall see cause) cast thereon water all about, and it will keepe the Trees moyst from tyme to tyme.

To cherishe Apple Trees.

If ye vse to throw (in Winter) all about pour Apple Trees on the rootes thereof, the Urine of olde men, or of stale pisse long kept, thei shall bring fruite much better, whiche is good for the Uine also, or if ye doe sprinkle or anopnt pour Apple tree rootes with the Gall of a Bull, thei shall beare the better.

To make an Apple growe in a Glasse.

To make an Apple grow within a Glasse, take a Glasse what fashion ye list, and put your Apple therein when he is but small, and binde hym fast to the Glasse, and the Glasse also to the Tree, and let hym growe, thus ye maie haue Apples of diuers proportions, accordyng to the fashion of pour Glasse. Thus maie ye make of Cucumbers, Gourdes, or Pomecitrons, the like fashion.

Thefe

Hese thꝛee bꝛaunches and figure of graffyng in the Shielde in Sommer is, the first bꝛaunche sheweth how the barke is taken of, the middle place sheweth, how it is set too, and the last bꝛaunche sheweth how to bindè hym on, in sauyng the oplet oꝛ eye from bꝛusing.

To graffe many sortes of Apples on one Tree.

YE maie graffe on one Apple tree at once, many kinde of Apples, as on euery bꝛaunche a contrarie fruite, as is afoꝛe declared, and of Peares the like: but see as nigh as ye can, that all your Cions be of like spꝛingyng, foꝛ els the one will not growe and shadowe the other.

To colour Apples.

TO haue coloured Apples, with what colour ye shall thinke good, ye shall boꝛe sloße a hole with an Auger, in the big= gest parte of the bodie of the Tree, vnto the middest thereof,

o? there aboutes, and then looke what colour ye will haue
them of. First ye shall take water, and mingle your colour
therewith, then stoppe it vp againe with a sho?te pinne made
of the same wood o? Tree, then waxe it round about:ye maie
mingle with the said colour what Spice ye list,to make them
taste thereafter: thus maie ye chaunge the colour and taste of
any Apple:Pour colours maie be of Saffron,Tourne soule,
B?asell,Sauders,o? other what ye shall see good.This must
bee doen befo?e the Sp?yng dooc come : Some doe saie,if ye
graffe on the Oliue stocke , o? on the Alder stocke , thei will
b?yng red Apples:Also thei saic,to graffe to haue fruite without co?e,ye shall graffe in both the ends of your Cion into the
stocke,and when thei be fast growne to the stocke,ye shall cut
it in the middest,and let the smaller ende growe vpwarde, o?
els take a Cion and graffe the small ende of the stocke doune-
warde,and so shall ye haue your Apple Tree on S. Lamberts
daie,(which is the seuenteene of September)thei shall neuer
waste,consume,no? waxe d?ie,whiche I doubt.

The setyng of Vine Plantes.

THese figures doe shew how ye ought to plant and set your
Uines,in two and two together , the one is haue a parte
of the olde Tree , and the other maie bee all of the last Cion:
but when ye plant hym with a parte of the olde Tree,he shall
commonly take roote the sooner then the newe Cion:ye must
weede them euery Moneth,and let not the yearth bee to close
aboue their rootes at the first , but now and then lose it with a
spade as ye shall see a raine past,fo? then thei shall enlarge,and
put fo?the better. Further herein ye shall vnderstande after.

How

How to proyne or cut a Vine in Winter.

Ƿis figure sheweth, how all Uines should be prop-
ned and cut, in a conuenient tyme after Christmas,
that when ye cut them, ye shall leaue his braunckes
very thinne, as ye see by this figure: ye shal neuer leaue aboue
twoo, or three leaders at the heade of any principall braunche,
ye must also cut them of in the middest betweene the knottes
of the young Cions, for those bee the leaders whiche will
bryng the Grape, the rest and order ye shall vnderstande as
followeth.

Of the Vine and Grape.

Omewhat I intende to speake of the orderyng of
the Uine and Grape, to plant or sett the Uine: the
Plantes or Settes whiche bee gathered from the
Uine (and so planted) are best, thei must not be old
gathered, nor lye long vnplanted after thei bee cutte, for then
thei will soone gather corruption, and when ye doe gather
your Plantes, ye must take heede to cutte and choose them,
whereas ye maie with the yong Cion, a ioynt of the old woed

with

with the newe, for the olde wood will sooner take roote then
the newe, and better to growe then if it were all pong Cion,
pe shall leaue the olde wood to the pong Cion, a foote or halfe
a foote, or a shaftment long, the poung Cion pe shall cutte the
length of three quarters of a parde or there aboutes, and pe
shall choose of those pong Cions that bee thickest iopnted, or
nigh iopntes togethers, and when pe shall plant or set them,
looke that pour ground be well digged in the Winter before,
then in Januarie pe maie bothe cutt and plant, but cut not in
the Frost, for that is daunger of all kinde of trees, or pe maie
plant in the beginnyng of Februarie, and when pe doe plant,
pe shall take two of those plantes, and set or laie them together, a foote deepe in the pearth, for two Plantes set together
will not so soone faile, as one alone, and laie them a foote long
wise in the pearth, so that there maie be aboue the earth three
or fower iopntes: pe maie plant a pong Cion with the olde, so
that he bee thicke or nigh iopnted, for then he is the better to
roote, and also to brpng fruite: then when pe haue set or laied
them in the pearth, then couer them well therewith, in treadyng it fast doune vnto the plantes, but let the endes of pour
Cions or Plantes bee turned vpright, aboue the pearth three
or fower iopntes, if there shall bee more when thei bee sett, pe
shall cut them of, and pe shall cut them alwaies in the middest
betweene the twoo iopntes, and then lett them so growe, and
see that pe weede them alwaies cleane, and once a Moneth
loose the earth rounde about them, and thei shall proue the better. If it bee verp drie and hot in the Sommer after, pe maie
water them, in makyng a hole with a crowe of Iron to the
roote, and there pe shall poore in water in the euenyng. As
for the proinyng of them is, when the Grape is taken and clu-
stered, then pe maie breake the next iopnt or twoo after the
Grape, of all suche superfluous Cions as pe shall see cause,
whiche will cause the Grape to waxe bigger: Pe maie also
breake awaie all superfluous buddes or slender braunches,
whiche commeth about the roote, or on the vnder braunches,
whiche pe thinke will haue no Grape, and when pe propne or

cut

cut them in Winter followyng, ye shall not cut the yong Cion nigh the olde, by thꝛee oꝛ fower iopntes, ye shall not cutte
them like Oꝛiars, to leaue a soꝛte of heads together on the
bꝛaunche, whiche doeth kill your Uine, ye shall leaue but one
head, oꝛ twoo at the moſt, of the yong Cions vppon the olde
bꝛaunche, and to cut thoſe yong Cions thꝛee oꝛ fower knottes
oꝛ iopntes of, foꝛ the young Cion doeth carie the Grape alwaies, and when ye leaue vpon a greate bꝛaunche many Cions, thei can not bee well nouriſhed, and after ye haue ſo cutte
them in Winter, ye shall binde them with Oꝛiars, in placing
thoſe yong bꝛaunches as ye shall ſee cauſe, and in the Spꝛyng
tyme, when the bꝛaunches are tender, ye shall binde them ſo,
that the ſtoꝛmie tempeſt oꝛ winde doe not hurte them, and to
binde them withall, the beſt is, greate ſoft Ruſhes, and when
the Grape is cluſtered, then ye maie bꝛeake of all ſuche bꝛaunches as is afoꝛe declared, vppon one olde bꝛaunche thꝛee oꝛ fower heads be enough, foꝛ the moꝛe heads your bꝛaunche hath,
the woꝛſe your Grape shall be nouriſhed, and when ye cutt of
any bꝛaunche, cut him of harde by oꝛ nigh the olde bꝛaunche: if
your Uine waxe olde, the beſt remedie is, if there growe any
yong Cion about the roote, ye shall in the Winter, cut of the
olde Uine harde by the grounde, oꝛ as nigh as ye can, and let
the poung Uine leade, and he will continue a long tyme, if ye
couer and fill the place about the roote with good earth again.
There is alſo vpon oꝛ by euery cluſter of Grapes, a ſmall Cion like a Pigges taile, turning about, which doeth take awaie
the ſappe from the Graps, if ye pinche it of harde by the ſtalke
of your Grape, your fruite shall be the greater. If your Uine
waxe to ranke and thick of bꝛaunches, ye shall digge the roote
in Winter and open the earth, and fill it vp againe with Sand
and Aſhes blende together, and whereas a Uine is vnfruitfull
and doeth not beare, ye shall boꝛe a hole (with an Auger) vnto
the harte oꝛ pith, in the bodie oꝛ thickeſt part thereof, then put
in the ſaied hole a ſmall ſtone, but fill not the hole cloſe therewith, but ſo that the ſickneſſe of the Uine maie paſſe thereby.
Then laie all aboute the roote of good yearth mingled with

R.iij. good

good dung, and so shall he not be vnfruictfull, but beare well e-
uer after: oz also, to cast of olde mens Urine oz pisse, all aboute
the roote of the barren Uine, and if he were halfe lost oz mard,
he should growe againe and waxe fruictfull as befoze: This is
to be doen in Winter.

¶ To haue Grapes without stones.

FOR to haue Grapes without stones, ye shall take yong
plantes oz bzaunches, and shall sette oz plant the toppe
oz small ende doune ward in the pearth, and so ye maie
sette twoo of them togethers foz failyng, as I haue afoze declared of the others, and those bzaunches shall bzyng Grapes
without stones.

¶ To make your Vine to bryng a Grape to taste like Claret.

TO make pour Uine to haue a Grape, to tast like Claret. Uine, and pleasaunt withall: ye shall boze a hole
in the stocke vnto the harte, oz pithe thereof, then shall
ye make a Lectuarie with the pouder of Cloues, of Cinamon
mingled with a little Fountaine oz running water, and fill
the saied hole therewith, and stoppe it faste and close with
Waxe, and so binde it fast thereon with a Linnen clothe, and
those Grapes shall taste like Claret wine.

¶ Of gatheryng your Grapes.

ALL Grapes that menne doe cutte, befoze thei are
through ripe, the Wine shall not bee naturall, noz
yet shall long endure good: But if ye will cutte oz
gather Grapes to haue them good, and to haue good Wine
thereof, ye shall cutte them in the full, oz sone after the full of
the Moone, when she is in Cancer, in Leo, in Scorpio, and in
Aquarius, the Moone beeyng in the waine, and vnder the
pearth.

¶ To knowe if your Grape bee ripe inough.

FOR to knowe if your Grape bee ripe inough, oz not,
whiche ye shall not onely knowe in the taste, but in sight
and taste together, as in taste if thei bee sweete, and full

in

in eatyng, and in ſight, if the ſtone will ſone fall out, bееying
chafed or bruſed, whiche is the beſte knowledge, and alſo whe=
ther thei bee white or blewe, it is all one matter: The good
Grape is he, whiche commeth out all watrie, or thoſe whiche
bee all clammie as Birdlime: By theſe ſignes ſhall ye knowe
when to cutte, beyng through ripe or not, and whereas you
doe preſſe your Wine, ye muſt make your place ſweete and
cleane, and your veſſelles within to bee cleane alſo, and ſe that
thei haue ſtrong heddes, and thoſe perſones whiche doe preſſe
the Grape, muſt looke their handes, feete, and bodie be cleane
waſhed, when as thei goe to preſſe the Grape, and that no
woman bee there hauyng her termes: And alſo ye ſhall eate of
no Cheboles, Scalions, Onions, or Garlike, Anniſſecdes,
or ſuche like : For all ſtrong ſauours your Wine will drawe
the infection thereof, and as ſone as your Grape is cutte and
gathered, you ſhall preſſe your Wine after as ſone as ye may,
whiche will make your Wine to be more pleaſaunt and ſtrō=
ger, for the Grapes whiche tarieth long vnpreſt, maketh the
Wine to bee ſmall and ill: ye muſt ſee that your veſſelles bee
newe, and ſweete within, and to bee waſhed with ſweete wa=
ter, and then well dried againe, and to perfume theim with
Maſticke, and ſuche ſweete vapour, and if your veſſell chance
not to bee ſweete, then ſhall ye Pitche hym on the ſides,
whiche Pitche will take awaie all euill, and ſuche ſtinkyng
ſauour therein.

¶ To proue or taſte Wine.

AND whenſoeuer ye will proue, or taſte any Wine,
the beſte tyme is, earely in the Mornyng, and take
with you three or fower ſoppes of bread, then dippe
one after another into the Wine, for therein ye ſhall finde (if
there be any) ſharpe taſte of the Wine. Thus I leaue (at this
preſent) to ſpeake any further here of the Uine and Grape. If
this my ſimple labour be taken in good part (gentle Reader)
it ſhall the more hereafter encourage me, to ſet forthe an other
booke more at large, touchyng the Arte of Plantyng and
Graffyng, with other thynges neceſſarie to be knowne.

¶ Here

¶ *Here followeth the beſte tymes how to*
order, or chooſe, and to ſette or
Plant Hoppes.

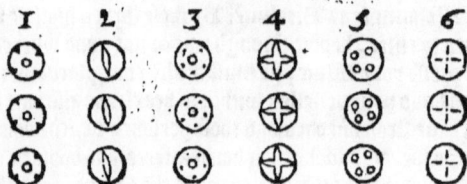

IN this figure ye ſhall vnderſtande, the placyng and makyng of the Hoppe hilles, by euery Sipher ouer his headde: The firſte place is
ſhewed, but one Pole ſette in the middes, and
the Hoppe beneath: The ſecond ſheweth, how
ſome doeth choppe downe a Spade in the middes of the Hille,
and therein layes his Hoppe rootes. The third place is ſhewed, how other ſome doe ſette out one Pole in the middes, and
the Hoppe rootes at holes put in rounde about. The fowerth
place ſheweth, how ſome choppes in a Spade croſſe in the
toppe, and there layes in his rootes. The fifte place ſheweth,
how ſome doe ſette fower Poles therein, and puttes the Hop
rounde about the Hill. The ſixte place ſheweth, that ſome vſe
to make croſſe holes in the ſides, and there layes in the Hoppe
rootes. Thus many practiſes haue been proued good: Proui-
ded alwaies, that your Hilles be of good fatte yearth, ſpecially in the middes downe vnto the bottome. This I thought ſufficient to ſhewe by this figure, the diuerſitie in ſettyng, whereof the laiyng of the Hoppe is counted the ſareſt waie.

THE beſte and common ſettyng tyme of Hoppes, is
from midde Nouember, to midde Februarie, then
muſt ye digge and cleanſe the grounde of weedes, and
mixe it well with good moulde and fatte yearth. Then deuide
your

your hilles a yarde one from an other orderly , in makyng
them a yarde a sunder, and two foote and a halfe broade in the
bottome, and when that ye plant them , ye shall laie in euery
hill three or fower rootes : Some doe in settyng of them laie
them croswise in the middest of the hill, and so couers them a-
gaine: some settes the rootes in fower parts of the hill, other-
some doe make holes rounde aboute the hilles, and puttes of
the rootes therein, and so couers them again light with earth:
of one shorte roote in a peare ye maie haue many plantes, to
set and laie as ye shall see it good , and it shall be sufficient for
euery plant , to haue twoo knottes within the grounde, and
one without: some doe choppe a Spade crosse in the hill, and
laies in crosse the Hoppe, and so couers it.

To choose your Hoppe.

YE shal choose your rootes best for your Hop, in the Sommer before ye shall plant them, for then ye shall see which
beares the Hop, for some there is that brynges none, but that
whiche beares, choose for your plants, and set of those in your
hilles, for so shal ye not be deceiued, and thei shal prosper well.

To sowe the seedes.

SOme doe holde, that ye maie sow among other seedes, the
seedes of Hops, and so will encrease and be good to set, or
els to make beds & sow them alone, wherby thei may increase
to be set, and when thei be strong, ye maie remoue and set them
in your hilles, and plant them as the other before mencioned.

The settyng your Poles.

THe best tyme is in Aprill, or when your rootes be sprong
halfe a yarde long or more, then by euery plant or Hoppe,
in your hilles, ye shall set vp a Pole of xiii. or xiiii. foote long, ·
or there aboutes, as cause shall require. Some doe vse to sett
but fower Poles in euery hill , whiche is thought sufficient,
and when ye shall sette them, see that ye sett them so fast that
greate windes doe not cast them doune. ·

How to proyne the Hoppe tree.

YE shall marke when the Hop doeth blossome, and knit in
the top, whiche shalbe perceiued to be the Hop, then take

and cut vp al the reſt growing thereaboutes(not hauing Hop thereon)hard by the earth, that all thoſe which carie the Hop, might be the better nouriſhed:thus ſhall ye doe in Sōmer,as ye ſhal ſee thē increaſe & grow,vntill the tyme of gatheryng.

To gather the Hoppe.

AT ſuche tyme aforꝛe Michelmaſſe as ye ſhal ſee your Hop waxe brꝛoune, oꝛ ſome what yellowe,then he is beſt to bee gathered in a drꝛie daie, in cuttyng your Hop by the grounde, then plucke vp your Pole therewith foꝛ ſhaking of your Hop, ſo carie them into ſome drꝛie houſe,and whē ye haue ſo pluckt them,ye ſhall laie them on boarded loftes,oꝛ on Hurdelles of clothes,that the winde maie drꝛie them, and the ayꝛe, but not in the Sunne , foꝛ the ſame will take awaie the ſtrength thereof, noꝛ with fire,foꝛ that will doe likewiſe,and ye ſhall daiely toſſe and turne them till thei be drꝛie:to trie them when thei are drꝛie,holo them in your hande a ſpace,and if thei cleaue together when ye open your hande,thei are not then drꝛie:but if thei ſhatter a ſunder in openyng your hande, then ye maie be ſure thei are drꝛie enough. If not,let them remaine,and vſe ye them as is beforꝛe ſaicd. Ye ſhall vnderſtand the drꝛineſſe of them is to pꝛeſerue them and long to laſt, but if neede bee, ye maie occupie them well vndrꝛied,with leſſe poꝛtion to ſowe.

What Poles are beſt.

YE ſhall prꝛepare your Poles of ſuche wood as is light and ſtiffe,and which will not bowe with euery winde, the beſt and meeteſt tyme to get them is in Winter, when the Sappe is gon doune,and as ſoone as ye haue taken of your Hop , laie your Poles in ſundrꝛie places vntill the next Spryng,whereby thei maie endure the longer.

How to order and dreſſe your Hilles.

AFter the firſt yere is paſt,your Hoppe being increaſed to moꝛe plentie of rootes in your hilles , ye ſhall after Michelmaſſe euery yere,open your hilles and caſt doune the tops vnto the rootes, vncoueryng them, and cut awaie all the ſu=perfluous roctes,ſome doeth plucke awaie all the rootes that ſprꝛeades abrꝛode without the hilles , then opens the hilles and puts

puts of good new earth vnto them, and so couers them again,
whiche shall keepe them from the Frost, and also make the
grounde fat, so shall ye let them remaine vnto the Spryng of
the yere, in Februarie or March, then againe if ye shall see a-
ny superfluous rootes, ye may take them awaie, and cut them
vp, and your Hoppe shall be the better, then againe call vp the
yearth about your hilles, and cleansing them from all weedes
and other rootes, whiche will take awaie their strength, if the
herbes remaine, so lett them rest till your Poles maie bee sett
therein.

Of grounde best for your Hoppe.

THE Hoppe delighteth and loueth a good and reasona-
ble fatte grounde, not very colde, nor yet to moyst, for
I haue seene them proue well in Flaunders, in drie
sandie fieldes, the Hoppe hilles beyng of good fatt yeerth, ye
maie (as some saie) for greate neede make your Hoppe grow
and beare on any kinde of rockie grounde, so that your hilles
be greate and fatte yearth, but the lower grounde commonly
proueth best, so that it stande well and hot in the Sunne.

A note of the rest aboue saied.

YE shall marke and vnderstande, all this order aboue
saied, is to haue many Hoppes and good, with a fewe
rootes and plantes placed in a small plotte of ground.
Ye shall vnderstande, the wilde Hoppe that groweth in the
Hedges, is as good to occupie as the other, to set or plant in a-
ny other place, but looke ye take not the barren Hop to plant,
some Hop will be barren for want of good yearth, and lacke
of dressyng, whiche ye shall perceiue (as I haue tolde you) in
the Sommer before, that when thei should beare thei will be
barren, whiche is for want of good fatt yearth, or an vnkinde
yere, or lacke of weeding and good orderyng. Therfore suche
as are mynded to bestowe labour on the grounde, maie haue as
good Hoppe growing in this Countrey, as is in other Coun-
treis: but if ye will not goe to the cost, to make Hop pardee, ye
maie with a light charge haue Hoppes growe in your Hedge
rowes, to serue as well as the other, and shall bee as good for

the quantitie as the other in all respectes: ye maie (for lacke of
grounde (plant Hoppe rootes in Hedge rowes, when ye doe
quicke sette vp Poles by them when tyme shall require in the
Spryng, and to bestowe euery Winter after the gatheryng
your Hoppe, on euery hill head, a shouell full of dung to com-
forte the yearth, for then will thei beare the more plentie of
Hoppe the next yeare followyng: to conclude, you that haue
groundes maie well practise in all thynges afore mencioned,
and specially to haue Hoppe in this orderyng, for your selues
and others: also ye shall giue encouragement for other to followe hereafter. I haue harde by credible persons, which haue
knowne a hundred hilles, (whiche is a small platt of ground)
to beare three hundred pounde of Hoppe, so that the commoditie is muche, and the gaines greate: and one pounde of our
Hoppe dried and ordered, will goe as farre as two pounde of
the best Hoppe that commeth from beyonde Seas. Thus
muche I thought meete and necessarie to write, of the orderyng and plantyng of the Hoppe.

How to packe your Hoppes.

When your Hops be well tossed and turned on boarded flowres, and well dryed (as I haue afore shewed) ye shall put them into greate Sackes accor-
ding to the quantitie of your Hoppes, and let them be troden
doune harde together, which will keepe their strength longer,
and so ye may reserue them, and take at your pleasure.
Some doe vse (whiche haue but small store) to
treade them into drie Fattes, and so reserue
them for their vse, whiche is counted
the better waie, and the lesse por-
tion doeth serue, and will
longer keepe their ver-
tue and strength.

Wishyng long life and prosperous health,
To all furtherers of this Commonwealth.

FINIS.

¶ Here followeth a neceſſary Table
(by Alphabete) to finde out quickly all ſe-
uerall particulars in this Booke afore mencioned,
by the numbers in this Table, ſeeking the like
number on the Pagine or leafe.

A

Buſhes

The Table.

Cuttyng

The Table.

Furni.

The Table.

Graffyng

The Table.

P.j.   Keepyng

The Table.

Ozchardes

.The Table.

P.ii.   Pzoy.

The Table.

Cions

The Table.

P.iij.   Trees

The Table.

FINIS.

THAT · BRINGETH · SVCH · LIGHT · WELCOM · THE · WIGHT

·I· ·W·

꒰Imprinted at London, for
Jhon VVight. 1 5 8 2.